Skills for Engineering and
Built Environment Students

Skills for Engineering and Built Environment Students: University to Career

John W. Davies

 macmillan education palgrave

First published 2016 by
PALGRAVE

Palgrave in the UK is an imprint of Macmillan Publishers Limited, registered in England, company number 785998, of 4 Crinan Street, London, N1 9XW.

Palgrave Macmillan in the US is a division of St Martin's Press LLC, 175 Fifth Avenue, New York, NY 10010.

Palgrave is a global imprint of the above companies and is represented throughout the world.

Palgrave® and Macmillan® are registered trademarks in the United States, the United Kingdom, Europe and other countries.

ISBN 978–1–137–40421–3 paperback

This book is printed on paper suitable for recycling and made from fully managed and sustained forest sources. Logging, pulping and manufacturing processes are expected to conform to the environmental regulations of the country of origin.

A catalogue record for this book is available from the British Library.

A catalog record for this book is available from the Library of Congress.

Printed in China

Contents

About the author

After graduating in civil engineering, John worked in the industry for eight years, becoming a chartered engineer. This was followed by over 30 years as an academic at the University of Westminster and at Coventry University, where for nine years he was Head of Department of Built Environment.

He is now Emeritus Professor of Civil Engineering at Coventry University and works as a part-time lecturer at Plymouth University (Civil Engineering) and at Petroc College of Further and Higher Education, Barnstaple (Built Environment).

Acknowledgements

I have talked to, listened to, quoted, and benefitted hugely from my contact with the following (identified as *Students*, *Graduates* and *Employers* where quoted).

Sam Bartle, George Berrett, Daniel Berridge, Jamie Bradley, Laura Bourne, Damian Clarke, Alun Howard, Geoff Hunt, Adam Lessiter, Conor Mitchell, Alexander Phillips, Rosemarie Poad, Ben Robbins, Inga Rowan, Lewis Safe, Arsalan Shafqat, Lewis Smith, Adam Tracy, Ed Yeates

Colleagues from various institutions have read my drafts and progressed the ideas far beyond anything I could have achieved by myself.

Coventry University: Steve Austin, Martin Beck, Erik Borg, Eoin Coakley, Tim Davis, Essie Ganjian, Irene Glendinning, Hal Igarashi, John Karadelis, Jon Ordidge, Geoff Powell, Messaoud Saidani, Neil Tsang, Mike Young

Plymouth University: Martin Borthwick, Andrew Fox, Kim Littlewood, Tamsin Turner

Petroc College of Further and Higher Education, Barnstaple: Alastair Green, Annemarie O'Brien, Tim Stapleton

Special thanks should also go to the anonymous reviewers.

Some colleagues have helped me in very specific ways:

Paul Smith, lecturer at Coventry University, without whom Chapter 10 would have far fewer valuable ideas;

Peter Evans, writer and writing coach, who has given me some great advice for Chapter 4;

Terry Noonan, employer-turned-lecturer (Swansea University), from whom I learned so much when we were student study colleagues, and who has provided some real insights into helping students prepare to be practitioners.

Paul Smith has kindly given me permission to use Figures 10.1 and 10.2.

John Karadelis has kindly given me permission to use Figure 14.6.

My family have been an inspiration: Ruth, Molly and Jack.

Thank you all so much.

Publisher's acknowledgements

The author and the publisher would like to thank the following organisations for permission to reproduce material from their publications:

Aalborg University Press, for permission to reproduce and adapt Table 11.1. From Austin S., Rutherford U. and Davies J.W. (2011) Large-scale integrated project for built environment undergraduate students: a case study. In: Davies, J.W., de Graaff, E. and Kolmos, A. (ed.) *PBL across the disciplines: research into best practice*. Aalborg University Press, 222–232.

International Journal of Engineering Education, for permission to reproduce the quotes in Box 11.1. From section 4.2 in Davies J.W. (2013) Using students with current industry experience to evaluate course delivery. *International Journal of Engineering Education*, 29 (5), 1199–1204.

John Karadelis, for permission to reproduce Figure 14.6.

Paul Smith, for permission to reproduce Figures 10.1 and 10.2.

Getting started

1

The initial transition to university

University study is all about transition, and we'll look at the various phases throughout this book. We start with the first transition – the one you experience at the start of your studies.

1.1 A change in your life

When he was 10, my son asked me, 'is it more difficult being a kid or a grown-up?' I thought it was a good question and deserved a good answer, so I thought as quickly as I could and said, 'It's about the same. When you're a kid the problem is everyone tells you what to do; and when you're a grown-up the problem is no one tells you what to do'.

I'm sure it's been a while since you thought of yourself as a kid, but the change takes a long time. Many people become students when this change is still having an impact, and at the same time something similar is happening to them educationally. It's a big transition.

Of course everyone is not the same. People become students at different stages in their lives. Not everyone has had the same previous education. Some people become university students straight from school, and some have been to college first. Some have significant experience of life, and of work, before starting a degree course. Some have been university students already, studying something different. And some are working while they're studying, as part-time students. Everyone faces slightly different challenges when they start a university course.

Straight from school

I've asked a lot of students who joined their course straight from school how they see the differences. Here's a summary; see if any of it rings true for you.

At school, teachers are on top of you, they know you well as an individual, and they nag you to complete work. At university, lecturers may be nice people, even approachable and friendly, responsive when you ask for help, but they are not 'looking out for you'. It's simply not their job. If you mess up it's your problem, not theirs. (In fact in one sense it is their problem, because their modules are monitored for pass rates.) But often the size of the cohort means that lecturers really can't look out for individual students. Of course another big difference is that most students at school are under 18 and therefore legally children, whereas most students at university are over 18 and therefore legally adults. Also at school you are likely to have contact (for learning anyway) with a limited number of teachers, and they will have clear roles. At university there may be several lecturers teaching even one module, and others, research students or teaching assistants, supporting your studies.

It's a teacher's job to look after you and to look over you, and they do this in a very professional way. During the school holidays recently my 12-year-old daughter dyed her hair with a dye that she thought would wash out easily. We happened to meet one of her teachers in the street, who laughed about how it would have to be out of her hair before they went back to school. It was all very jokey. But back at school apparently the same teacher, with a completely straight face, inspected her hair to make sure it complied with the school rules. (And it only just did, in spite of being washed five times a day for the last few days of the holidays.)

> *For me school was all about rules, 'not allowed to wear trainers after lunch break'; I hated it. But here, if you like rules you'll struggle.*   ♟ **STUDENT**

I can't do that stern-face thing – that's why I've never wanted to be a school teacher. I'd rather just be 'friendly'. But that's why my students can't use me to judge whether what they're doing is good or bad. And that applies to most other university lecturers. It's not an advantage or a disadvantage, but you need to be ready for it. The point is really that if you've gone to university straight from school, whether you realise it or not, and whether you like it or not, you've been used to having someone tell you whether what you are doing is basically OK or basically not OK – certainly teachers, and also, depending on your situation, your parents. Even though you don't like it, you might have got used to your parents asking: 'Have you got any homework?' Now, you are much more on your own.

> *Main difference between school and university? At university, if you don't do it, nobody notices.*   ♟ **FINAL-YEAR STUDENT**

At school your study outside the classroom is effectively managed by your teacher. At university it is not: work can build up quickly and suddenly; coursework deadlines can clash. Perhaps we could say that a 'school atmosphere' is where staff take responsibility for your learning; university education is based on students thinking for themselves, and managing their own time.

> *At university ... self-study – it's a jump, even bigger than I thought.*
> ⚇ **STUDENT**

At school teaching groups tend to be smaller than at university. You know everyone well. Another aspect is that at school some teachers know you as 'a whole person'. At university that is less common. I remember in my first year as a student joining a sports club, and being surprised during the first training session that there was no one like a school teacher around. I almost thought 'what's the point of doing this if there's no teacher to impress?' But at university you do things for yourself, not for your teachers. You study to do well in your qualification, partly because that is satisfying in its own right and partly because the qualification will have value for your career and your life. You play sport for pleasure, for fitness, for satisfaction; if it's a team sport, then of course you also strive for the good of the team. But you don't do these things because a teacher will notice (or, you might feel, because a teacher is 'checking up on you').

> *At school they may try to prepare you for university, but they don't.*
> ⚇ **STUDENT**

College first

If you went to college to do A-levels or a BTEC you'll be used to a bit more freedom – you can wear what you want, dye your hair, smoke (provided you use the designated shelter), without being told you're breaking the rules. But these are superficial things. Lecturers at college still look after you and look over you. At college, like school, you're likely to be in relatively small groups – you're 'known'. You're a little bit more independent but the atmosphere is still fairly cosy.

I've started teaching at a college recently, after a lot of experience of lecturing at university. I'm still surprised and impressed by the level of care and responsibility that staff show for their FE and A-level students. Emails are constantly being logged on the system, like 'I'm very concerned about Ollie's motivation and engagement. I've discussed it by phone with his father and we've agreed an approach of heightened monitoring'. Or 'Ollie was late again this morning. He knew it was an important revision session'. Then again, 'Ollie missed the mock test this morning. I have issued a formal warning'.

Part of me thinks 'poor Ollie, perhaps we ought to get off his back and just let him sort himself out'. If this was happening at university and, as a lecturer, I found myself talking to Ollie, I might try to find out if something is wrong – but that's about all. My inclination would be to let Ollie sort himself out. He might enjoy being left in peace, but that would only work if he took responsibility for himself in the end. The attention he is getting at college may easily be just what he needs. If he goes to university and assumes someone will still be constantly looking out for him, it could all go horribly wrong.

In some ways college is a good preparation for university, it gives you independence in many ways, but that doesn't generally apply to studying, and that's the most important thing.

Gap year

You'll have already learnt a lot about living independently; now the challenge is to apply that to studying.

Already seen a bit of life

If you've been away from education for a while you may feel anxious about getting back into studying, but you do have some advantages. You are likely to have more maturity than students straight from school or college; this is because you're a bit older but also because of your experience of life. I asked a group of more experienced full-time students what advantages they thought they had, and the main things they mentioned were focus and attitude. Some said that their experience of working in a dead-end job had made them appreciate the opportunity they had to get a qualification that would lead to a real profession. They were determined not to mess it up.

What motivates me is knowing how bad work can be!   🙎 **MATURE STUDENT**

Studying part-time

If you are studying part-time you may feel, quite rightly, that your situation is substantially different from that of full-timers. For many full-timers, the most important issues of transition may relate to the significant changes in their lives: moving away from home and being given greatly increased independence in the way they live and how they approach their studies. As a part-time student you are probably experiencing fewer changes in your life: your home environment may easily be continuing as before, even providing continuity, although it may now include additional competition for personal time in the form of studying.

However, part-time students face other sorts of challenges: shifting priorities between work, life and study; additional time and academic pressures; possibly the stresses of re-adjusting to studying.

If you are employed in a professional discipline related to your studies you have the potential to benefit from the juxtaposition of academic learning and its practical applications. This provides a really excellent context for your education.

> *I could see that the work I was doing at university was directly related to what I was then doing the next day in the workplace for real.*
> 👤 **GRADUATE (PART-TIME COURSE)**

You will also be able to apply professional skills acquired at work along with the distinctive 'workplace attitude' to studies that most part-timers have, which generally enhances their academic achievements.

> *Going to work gives you responsibility and a useful attitude, how to manage your time.* 👤 **PART-TIME STUDENT**

If you are employed in a different area from your study discipline you may experience the second benefit but not the first. In any case, studying must compete for time with work and family/leisure commitments.

Sometimes part-time students feel they are at a disadvantage in some subjects, most commonly maths, but just about every part-time student I have known has had the motivation to overcome this. There will be help available – you just need to find it and make the most of it.

1.2 What you have to get used to

This is a sort of summary of what's just been said, but it applies to some extent to everyone, regardless of previous educational experience.

May sound easy	May sound difficult
Independence	You're on your own
Freedom	You've got to function without a 'structure'
No rules	No rules
Help is always available	You've got to ask for it
There's no one constantly 'on your case'	You've got to keep yourself on track

It's not a hostile environment – quite the opposite in fact. Most academic staff are friendly, relaxed in a way, and helpful if you ask for help. But that's the point: you've got to ask, you've got to know that you need help, and go and find it. In that sense you're on your own, you've got to look after yourself. So ... not at all hostile, but you're on your own. That's what you've got to grasp.   ⚇ **GRADUATE**

In my first year I didn't know how to perform at my best without a structure. I was uncomfortable with the 'freedom' [of being a student], which I saw as chaotic, as a distraction. Toward the end of my degree I had learned how to use the freedom to perform at my best, to prioritise as I wanted, without having to follow the irrelevant aspects of someone else's structure.   ⚇ **GRADUATE**

1.3   Motivation

Why have you started a university degree? Here are some possible reasons.

1   Because you did well in A-levels, BTEC or whatever
2   Because your parents want you to
3   Because your friends are doing it
4   Because your teachers advised you to
5   Because you've got to be a student some time in your life
6   Because you want to learn and you think the course will be interesting
7   Because you want a good job at the end of it
8   Because you want an interesting career

In section 1.1 you may only have read the part that you felt applied to you, 'straight from school' etc. It's true that the transition to studying at university is very different, depending on your background. Let's think now about how your background affects your motivation.

    I went to university straight from school. Did I think carefully and have a clear reason?

1 not particularly (I did well enough at school)
2 definitely (thought too much, really)
3, 4 and 5 quite a bit
6 didn't think much about it
7 and 8 sort of taken for granted: I had 'chosen' a career path and that was why I started a degree in that discipline, but it didn't mean anything very specific to me.

    So what is *your* motivation?

It shouldn't be about what other people want or expect you to do. You may be thinking, 'OK, I understand that, it should be about what I want to do – but that's the problem'. Let's clarify what you should be identifying here. It's not what you want to do this week or even this year; although at the same time no one's expecting you to answer the question 'what do I want to do for the rest of my life?' It's really about direction – what direction do *you* want to take? In particular it's about whether an academic qualification in the field you have chosen will help you move in the right direction, or help you find your direction.

In some subjects there is a special admiration for student motivation that is based purely on a desire to learn (6. in the list). There is even a term for it – 'intrinsic motivation'. This is in contrast to 'extrinsic motivation' which is all about seeking rewards, in the case of university study: good marks and a good degree. For professionally-recognised qualifications like those in engineering and built environment this also includes getting a good job at the end of it.

The benefits of a vocational degree in terms of the profession that follows are understood by most students in those disciplines, but are probably felt particularly strongly by people who have had a period outside education before starting their course.

> You know why you're coming here, you know what's at the end of it – if you do well. 𐐥 **MATURE STUDENT**

I think this is an excellent type of motivation. Students who've seen a bit of life, and certainly students who are studying part-time, have heaps of it, and it improves their performance and keeps them going.

We'll now look a little more deeply at how your motivation can see you through the challenges you have taken on. Of course if you've chosen a subject you really like, all these challenges should be easier to overcome. It makes sense to divide this into two sections, 1.4 on engineering and 1.5 on built environment. There are a few final thoughts relevant to both subject areas in 1.6.

1.4 Studying engineering

Definition of engineering

The word 'engineer' comes from the Latin word *ingenium* meaning cleverness, or natural ability. Engineering is about being ingenious, and about using ingenuity to solve problems. Of course many other people solve problems for a profession, but for engineers the problems tend to be practical and the solutions tend to be technical. Blockley (2012) says 'Engineering is the discipline of using scientific and technical knowledge to imagine, design, create, make, operate, maintain and dismantle complex devices, machines, structures, systems, and processes that support human endeavour'.

Myths and truths about studying engineering

At the start you won't be sure what to expect on an engineering course. At some stage early on you may have a bad moment and think 'is this really for me?' Here are some myths and truths about studying engineering.

> *It's harder work than other courses*
> *You've got to be good at maths*
> *It's hard for women to progress, however good they are*
> *It's for people who are practical, who like fixing motorbikes etc.*
> *It's not creative*
> *It's not about helping people*
> *At least you can be sure of getting a job at the end of it*

Let's look at these one by one.

It's harder work than other courses
Engineering courses tend to have full timetables, with some early starts and late finishes. There also tends to be a high loading of coursework, though some universities are reducing this. So in these respects it is harder work than some subjects, though there's pressure on universities to increase class time for all courses (to convince parents that the fees are good value for money). So stories of English students attending two lectures a week may be a thing of the past. Anyway, you should talk to English students – you may find that the amount of reading they are expected to do would be too much for you. Work doesn't seem so hard if it's your sort of work and it's interesting.

You've got to be good at maths
This is an interesting one. You definitely don't have to be a genius at pure maths to be a successful engineering student. But you can't afford to be bad at maths either. People who drop maths at 16 and later decide to study engineering can be at a big disadvantage. If they aren't confident about basic algebraic manipulation, for example, they can really struggle in the core engineering subjects. It's a great shame – it's the maths that is letting them down, not their understanding of the engineering; but they've got to get on top of this lack of confidence. Engineers quantify: the way they solve problems often involves quantification. So basics maths, at least, must be a comfort zone.

It's hard for women to progress, however good they are
Women are under-represented in the engineering profession and on engineering courses. A lot of work is being done to try to change this. Some women say they were not particularly encouraged at school to pursue their interest in

engineering, though others say they were. When women start an engineering course they have probably thought about their choice at least as hard as male students, and if anything that puts them at an advantage. During their studies, unless they are very unlucky, they should receive nothing but encouragement to succeed – and they are very likely to do at least as well as their male colleagues.

It's for people who are practical, who like fixing motorbikes etc.
This was something that really bothered me when I was thinking of doing engineering. I told my father I was worried about the fact that I wasn't remotely interested in fixing motorbikes or doing up old cars. Later he told me delightedly that he had been talking to a friend (who was a engineer) about my plans, and the friend, without prompting, said 'I hope that's because he's good at maths not because he likes tinkering with motorbikes'. In fact I wasn't too sure about the first part, but I was very relieved about the second. The work of engineers usually involves some practical outcome, but engineers are fundamentally thinkers, problem-solvers; they do their work with their brains.

It's not creative
Well, engineering is not 'artistic', not an opportunity for purely personal self-expression. There are plenty of courses that are, of course, though very few graduates make a living from their art. But engineering is about creating – creating solutions, and very often creating 'things'. These are things that weren't there before, often giving the engineer a huge sense of achievement. And by the way, if you are creative in the sense of writing poetry or playing the guitar or anything else – great. Make sure that remains important in your life. It's no reason to feel that you won't enjoy being an engineer. There is no contradiction, no conflict; you can be both. (There's more about this shortly.)

It's not about helping people
I'll be more direct this time: it *is* about helping people. That is the main purpose of engineering. But unlike nurses or a social workers, engineers do not usually have day-to-day direct contact with the people they are helping.

At least you can be sure of getting a job at the end of it
There are lots of good jobs in engineering and lots of evidence that there aren't enough qualified engineers in the world. Of course for you to get a good job, an employer must want to employ you (you as an individual, not just as a holder of a qualification). Putting yourself in the best position to get the best job when you graduate is still important (and that's what a lot of this book is about).

A 'typical' engineering student?

If you ever have doubts about your choice of degree course don't get drawn into comparing yourself with a stereotype of an engineering student. Engineering is a very wide field, even within each sub-discipline, and it needs a wide range of people: problem-solvers, designers, number-crunchers, computer modellers, organisers, managers, leaders, environmentalists, academics, entrepreneurs, negotiators, creative people, business people, outdoor people, people people, people who want to make the world a better place, who want to improve people's lives, who want to make money, people who like big challenges, like intricate challenges, like solving physical problems, like solving abstract problems...

And, as I have already mentioned, if you have a very strong interest outside engineering, don't make that, in itself, a reason to question being an engineering student. I have known practising engineers who have also been brilliant in music or sport. To advance your passion, whatever it may be, what matters is what you do, not what you don't do.

1.5 Studying built environment

Whereas students in all branches of engineering tend to start on their course with fairly similar qualifications but leave to go into a wide variety of industries, in contrast built environment students come on to the wide range of courses with very different skills and aspirations, but all become qualified potentially to join the same industry: that involved with the planning, conceiving, promoting, designing, constructing and managing of the built environment. Good built environment courses provide opportunities for students to work with colleagues in the other disciplines, but there are very strongly defined and distinct approaches to education in (for example) architecture, planning and surveying. These are reinforced by the separate professions and their involvement in the professional accreditation of courses.

So it's not easy to generalise about studying built environment, and it's probably best to consider different sectors of built environment separately. (Civil engineering and building services engineering are covered by 'engineering' in the previous section.) You've already made a choice, and you've probably put a lot of thought into that. But don't be worried if you have 'second thoughts' – that's only natural. Hopefully some of what follows may be useful. All the disciplines overlap, so any of it may be relevant to you.

Design-based disciplines

Many different professions include design work, but I'm thinking here of the professional built environment areas where design is at the core, including **architecture, architectural technology** and **landscape architecture**.

Often the motivation for joining this type of course is a desire to be creative. When you have thought about the question 'what are you really passionate about?' you may have thought straightaway: 'being creative'. If beautiful structures excite you, then a course like architecture is a logical choice. And all the way into your course, creativity may have remained the focus, for example in the emphasis given by the tutor who looked at your portfolio when you went for interview.

In fact the range of the course will be incredibly wide, from philosophy to understanding structural behaviour. That range makes the study experience very rich – but it is not all about being creative. The balance between arts and science is different on different courses. You may have taken this into account when you chose your course. But some students start a course that has the wrong balance for them, either because they have misunderstood the wide requirements of a design-based course or because they have been unlucky in their choice of course.

Some students are utterly confused at first: the course activities seem a long way from designing buildings. What's happening may be creative, but in an abstract or philosophical way which comes as a complete surprise.

You will need to be able to present your ideas in more than one way and will need to develop skills in a number of areas, including:

- freehand sketching
- model-making
- photography/film/montage
- computer-aided drafting
- painting

There will be facilities to support you in these areas but you will need to learn for yourself.

You also need to be ready for the fact that your work may be exposed to some quite sharp criticism, sometimes in front of others. The important thing is to know how to learn from this. Some students who have been used to doing very well at school or college are surprised when their work is criticised in this way, and may take a while to adjust to different types of expectation. The process may be intimidating at first, but the key is to realise that it's not about you, it's about your work.

A lot of the activity comes in the form of big projects with fixed deadlines. Of course the right response is to plan your work carefully, but in reality it may come down to finishing your work with a significant lack of sleep (and students on design-based courses seem curiously proud of this).

Surveying

We're talking about a range of disciplines here, mostly with professional recognition via the RICS (Royal Institution of Chartered Surveyors). Two of the most popular are **building surveying** and **quantity surveying**. It is often said that an interest in these areas comes via friends or family: you find out most easily what a building surveyor does from someone who does that job themselves.

I'm sure you wouldn't have chosen your course if you didn't have some idea of where it would lead, but if you don't have this first-hand contact via friends or family it would certainly pay to find out as much as you can about the profession early on in your course. Both building surveying and quantity surveying lead to a wide range of careers, but at the same time they suit people with distinctive interests and qualities. Course content has a wide range, from science and structures at one end to economics and management at the other, with lots about building fabric and costs in between.

> *Subject content is very broad; students need to have an open mind and be flexible.* 👤 **LECTURER**

Content is also broad because built environment is a team industry and you will inevitably work across the disciplines. So you will need to learn about other professions, and you may even have common studies with other built environment disciplines in year 1, as discussed further under 'Management'.

Management

Course titles like **Construction Management**, **Project Management** or **Facilities Management** convey a fairly clear idea of course content and potential career route. The focus is on planning, coordinating and managing the construction process, and managing facilities once constructed. In the job you will need to be a good team worker, good at creative problem-solving, and good at organising and planning. However, if you are coming on to your course straight from A-levels you will have very little preparation academically for the subject content. It is often practical experience that sustains interest in these areas, and it's a good idea to seek this out if you don't already have it.

The content of your course will be wide-ranging especially if you are on a course where there is a high degree of commonality in year 1 between courses in architectural technology, building surveying, quantity surveying and construction management. This commonality is a good thing even if it means you have to master a wide range of new subjects; it means you will have knowledge of other professional areas, and it may even give opportunities for course transfer.

You may be surprised by the emphasis given on your course to dealing with people. But really you shouldn't be; after all you have chosen a course that has the word 'management' in the title.

> *Dealing with people – that's the key area in construction management.*
> ♟ **LECTURER**

Planning, development and housing

You are likely to have chosen your course because of a strong interest in people, their environment and the way they live their lives. There are big issues here, and you may have started engaging with them because of your interest in your local community. In preparing for a role which involves balancing social, economic and environmental factors, part of the challenge is the need for a broad perspective, and mastery of a range of complex topics as you progress on your course. The sheer range of subjects can be a challenge; learning in these areas involves the synthesis of disparate parts. This includes what may seem at first to be a big span between theory and practice. It should all tie up at the end, but it can be hard to see this at the start of the course.

You will have the opportunity to specialise – there is a wide range of career paths. Try to find out what interests you in the early stages of your course, in terms of theoretical and practice aspects.

Your motivation is likely to be sustained by commitments outside the course and the professional area, for example involvement in your local community, or a strong interest in politics, or global issues. This is just as it should be: your breadth of vision is needed.

1.6   Getting through the initial transition

This chapter has considered the initial transition when you start a university course. We have looked at it from a number of perspectives. Your own experience of the transition will clearly depend on what you have been doing before you started your course, and on your motivation for studying a degree. The discipline you have chosen will obviously make a difference too. So what's the best advice overall? Here are my recommendations.

1   Realise that everyone has a slightly different experience of this transition. Think about your individual situation: what is changing most for you, what motivates you, where you want to go. If you feel any uncertainty at the start of your course that's only natural; lots of people do.

2 Take every opportunity to find out more about the subject you have chosen. Remember what attracted you to it, and make that part of your motivation. Get involved in introductory sessions on the course, attend events, immerse yourself in the discipline.

3 Adjust as smoothly as you can to your new environment. If you've come straight from school, even from college, realise that now you have to make your own decisions and take responsibility for your studies. You probably will need to develop a new way of approaching things. Read on – this is what we cover in Chapter 2.

2

Organisation, initiative and flexibility

This chapter is about being in the right state of mind for higher education, and coping with the real challenges. Some of the challenges will be academic, but the greatest challenges are likely to be in personal organisation, in taking the initiative, and in adapting to unfamiliar demands.

> *We have this forum where staff here meet up occasionally with local school teachers. The teachers all seem to think the main challenges when students come to university are academic. They talk about 'stepping up academically'. Most of us here think the main challenges, in the first year at least, are about organisation and maintaining motivation. Our students need to be able to perform consistently across all modules, and to ask for the help they need, without a teacher looking after them all the time.* 𐎦 **LECTURER**

2.1   Eight principles

The keywords in this chapter are organisation, initiative and flexibility. Here are eight principles for being successful as you adjust to higher education.

I know – Chapter 1 was all about thinking for yourself and not expecting to be looked after, and now here's me telling you what to do. But these are principles, not instructions. You can use them how you wish.

1   Make your own decisions

Of course you'll have colleagues, friends, study companions. There's no suggestion that you should turn study into a lonely occupation, but you must make decisions about your studies for yourself. The more patronising way of putting this is 'take responsibility for your studies'. Don't do something because everybody else is doing it. Don't wait to find out what everyone else thinks before making up your mind. Decide what's important for yourself. Here's a specific example: don't wait for other people to remind you about deadlines.

2   Establish your individual identity

In many situations on your course it feels like you are lost in a crowd – in large lectures, on a crowded campus. But you are unique. That is not some cliché aimed at bolstering your confidence. It's true in a very practical sense – the scores you get in assessments, the precise nature of your achievements, your eventual career path, will all be unique. You are not just unique in the end, you are unique from the beginning. Don't make the mistake of assuming that, apart from a few wild extraverts, students are anonymous to their departments and lecturers. Through your engagement with your course and your classes, staff will come to know you as unique. And that's important: you are likely to need at least one of your lecturers to write a reference about you at some stage. To be a good reference it will require the lecturer to know you as an individual. Talking about your 'profile' may seem pompous to you while you're a student, maybe something for later in your career, but your profile is your uniqueness, and it is enhanced by everything you do that is positive and committed.

3   Think about life and study

Education is not just about learning subjects, it is about life and how to live a better life. Also you came to university to be a student, and everyone knows about the 'student life'. There's a lot to do that isn't studying! It would be poor advice, and pointless, to say 'just concentrate on your studies'. This is your chance to broaden your horizons, meet people, think, discover things about the world and about yourself, throw yourself into your interests, discover new interests, break the record, write songs, win the cup...

> *It's very difficult to get the study-life balance right, but you've got to get it right, negotiate your way. I could say 'don't do this, don't do that', but it's all part of the student experience. You've got to find a balance.* ⛈ **RECENT GRADUATE**

4   Organise and plan

All engineers and built environment professionals need some project-management skills and you are almost certain to start acquiring them as part of your course. But here we are talking about the management skills that you apply to your own project – that of completing your course successfully. There are the usual components: time (mostly yours), resources (you), plant/facilities (computer, study space), milestones (coursework deadlines, exams), the need to prioritise. No one's expecting you to draw a time chart for every assignment, but you should pay attention to the need to plan. You may be managing big projects when you're a professional so it certainly won't do any harm to show you can manage your own time while you're a student. One obvious thing to look out for is bunching of coursework deadlines; this will call for sequencing and prioritisation. Where time is short it should be distributed between assignments, taking into account their relative importance. What time and date are you aiming for? Most universities are very strict on coursework submission. If the hand-in deadline is 3 pm on Thursday, how long (working backwards) will it take to finalise and bind, how long to print out (allowing for printer queues, printer failures), how long to check before printing…? By the way, my college students are surprised when I tell them how strictly coursework submission deadlines are applied at university. If something goes wrong you may be able to make a formal application for special circumstances to be taken into account, but you can't just go up to someone at the last minute and say 'please can I hand my work in tomorrow because I left my notes on the bus?' You've got to get used to planning everything carefully.

Day	1	2	3	4	5	6	7	8	9	10	11	12
Analyse brief	█											
Background reading		█	█									
Investigate options				█	█							
Write draft report							█	█				
Improve and finalise									█			
Final check and print										█		
Finalise and bind											█	
Submission												█

This is a Gantt chart, indicating start and finish times of activities. It can be a simple table (like this) or it can be produced using project management software. Gantt charts can help you plan simple jobs, or complex projects.

5 Keep an eye on where it's leading

As we've been discussing in Chapter 1, for some students the first few weeks and months on a university course can be a big transition. But in fact the course itself is a transition all the way through. If you maintain your interest in your discipline, and you get a graduate job at the end, the overall transition will have been from school-kid, college-kid, or whatever, to career professional. I suppose there are two good reasons for keeping an eye on where your course is leading: to help you get a good job at the end, and to make your course more interesting all the way through (by providing motivation). You can easily satisfy your curiosity about what's going on in your field and the types of jobs that are out there by looking online, in magazines, or in material provided by the university's employability/careers team. I'm not saying you need to become obsessed about career planning. If you're a full-time student you should be 100 per cent a student, expanding your horizons and getting involved in student life. But keep up your interest in your discipline generally, and keep an eye on where your studies are leading. Of course the best way to find out about practice in your profession is to experience it first-hand. We're going to look in Chapter 16 at the benefits of work experience and how to find it. But just for now it's worth reflecting that at some stage during your course you are likely to be looking for work experience. By then you will need to have some idea of what you want. In your interview you will probably be asked 'what interests you within our profession?' Your interviewer will find out a lot about you from your answer.

6   Take opportunities

As we discussed in Chapter 1 there are no teachers or college lecturers (or even parents) looking out for you, guiding you to do the right thing. And as we discussed, that means you have to look after yourself, make decisions for yourself. Opportunities are a big thing at university. In fact you are surrounded by opportunities, so much so that sometimes the problem is making choices about opportunities, maybe even rejecting some in favour of others. You are bombarded by messages, posters, information generally. Some of these opportunities – social, sport, entertainment – are tempting and may need balancing. Then there are the more serious, course-related opportunities: lectures by engineers, visits, professional institution meetings, the opportunity to be mentored by a practising professional. These opportunities really matter.

> *Do as much networking as you can – attend professional institution events. Some students don't even think about it – they think it's for later in their career – but no, it's the opposite. Go to events, talk to people: it builds up your confidence, you can meet very interesting people.*   & **FINAL-YEAR STUDENT**

   You should be prepared to take the initiative yourself, not just accept the opportunities that come your way. This means being active in pursuing and even creating opportunities.

7 Be flexible

There is a big transition, there are opportunities, and the need to take the initiative. You need to adapt, to be flexible, prepared to change. Look beyond what you've been used to doing in the past. Welcome the idea that you will change, and that you will develop as a person. Try things; then reflect and improve. You can do that as a student far more than you could as a school kid, and far more than you will be able to as a professional.

8 Be yourself – but think about what that is

Flexible, yes, but you must be yourself. This means knowing yourself, knowing your strengths and weaknesses. It also means deciding what you want: what is important to you, what motivates you. If anyone connected with your course or career – colleague, lecturer, employer – asks you what you want, don't tell them what you think they want to hear. They want to know what *you* want. Justifying what you want can be hard with people who are older than you, but you know what you want and they don't. Often people imagine there is pressure to follow a conventional path when in reality there isn't. You may have to do things in your course or career that you don't enjoy at the time, but try to see the bigger picture. Find the path that suits you.

2.2   What my students say: secrets of success

I've asked a lot of students what they think are the secrets of success. Here are some highlights. I'm trying to use their words as much as possible. There's some serious stuff here!

Focus	Focus on what you're doing – stick at it – avoid massive distractions.
Colleagues	Associate with other people with focus.
Strengths	Know your strengths but don't make them a comfort zone; develop strategies to overcome your weaknesses.
Work	Don't be lazy. Don't underestimate work. If you find you've got lots of free time you're doing something wrong. Students are here because they want to be, and they want to succeed; the work ethic is strong.
Team work	Team with people who complement your strengths and weaknesses. If you're put in a group, identify each other's strengths and weaknesses.

Problem-solving	Follow instructions – interrogate the brief. Don't overthink – look for focused solutions.
Help	Ask for help. Lecturers are not scary. Ask as many questions as you need to, especially about assignments. Realise that you may not get guidance unless you ask for it.
Adapting	Adapt to different teaching styles.
Skills	Get your maths right. If you think you need to, work on your literacy skills.
Planning	For the first few weeks you are sort of told what to do; then you're on your own. Think about time management. Keep up to date. Realise work can build up quickly. Deadlines all come together. No one plans for you. You can't leave things to the last minute.
Organisation	Always set out notes and calculations clearly. Organise your material.
Environment	Have somewhere to study that's quiet, with a suitable layout.
Health	Have a healthy lifestyle; think about your diet. Exercise is really important: helps sleep, helps concentration.
Balance	Manage the life/study balance (social/physical/study/part-time work). Sports, interests: get involved; but think about the life/study balance.

Study skills

3
Basic academic skills

You've got this far in your academic career so you must have academic skills. You've passed exams, probably written plenty of successful assignments, and learned from your course experience. This chapter picks out some aspects of study at university that you may not be so familiar with, and that may take a bit of getting used to. We've already considered one of these in Chapter 1 (1.4 and 1.5), the fact that you are now concentrating on a particular vocational area, with its own characteristics and demands.

Basic academic skills are mostly about learning; and assessment is about showing that you have learned. So this chapter is divided into two parts:

- Who and what do we learn from?
- How do we show that we have learned something?

Notice that the first question is not 'How do we learn?' That's a much harder question which we will return to a few times (in Chapter 12 especially).

3.1   Who and what do we learn from?

It seems helpful to divide the answer into three areas. Thinking specifically about the subject matter of our degree course, we learn from:

- people
- experiences
- sources

Expanding that a bit, we might end up with:

Who and what do we learn from?			
	People	*Experiences*	*Sources*
Directly related to the course	Contact with staff Fellow students	Classes Study and practice Assessment and feedback	VLE* Printed material provided Internet Books, documents
Wider	Friends	Previous education Work experiences Life	All media

*VLE: virtual learning environment, your university's system for making materials and opportunities for learning available via the web

By 'wider' I mean wider sources of learning about our subject specifically. As an example, we might learn something very relevant to our subject from a story in the news. But in this chapter we will concentrate on aspects directly related to the course. Let's look at these in more detail using the categories in the table.

Contact with staff

The first thing to note is that your lecturers teach you and they assess you. When I first started lecturing, that dual role used to bother me. But I quickly became more comfortable with it and I think most lecturers are the same. I suppose we think, 'If the students are OK with the material they'll pass, and if they're not they won't'. No one really wants a course where people pass when they shouldn't. But I can understand that students are always conscious of their lecturers' dual role. When I'm enjoying some nice social time with students, I notice that occasionally someone can't resist the temptation to make a joke about increasing their chances of passing the module. But lecturers have to separate out their personal opinions from their responsibility to maintain the standard of the course.

You may enjoy the teaching style of some lecturers more than others, but the most important thing is that your lecturers are experts in their subject area, either through research or practice (see Chapter 12). You should concentrate on benefitting from their expertise, by listening, questioning and interacting when you can. Get to know the staff, remember their names and roles.

Other people who support your classes – teaching assistants, research students, lab technicians – may have a lot to offer in terms of expertise too. If a researcher is supporting a class it is worth finding out what their specialisation is – they'll probably appreciate being asked. They should still be able to help you in the class even if they are working outside their specialist area, although if you feel they are not helping, ask someone else.

Communicating with lecturers: some tips

1 When you are using email, write in a 'polite' style, not in the style you might use to text a friend.
2 Don't be impatient about receiving a reply. You will be a priority in your lecturer's mind, but there may be others too.
3 If you arrange a meeting, bring specific queries.

By the way, how should you address your lecturers? This includes addressing them in speech or in writing. I have taught some students who I'm guessing attended international private schools and addressed me as 'sir'. I find this excruciatingly inappropriate. Of course there is no simple answer to the question; and it's a decision you have to make in many other situations too. Like so many things at university, you've got to decide for yourself. A safe starting point is [Title] [Surname], but be careful not to say Mr Smith when it should be Dr Smith or Professor Smith. In many cases it is normal and expected that you would use their first name.

Fellow students

There's nothing better than sharing understanding between colleagues. You may understand something best if a fellow student explains it. And if you are able to help someone else, it is great for your self-confidence. Apart from all the informal support, many universities have structured peer-mentoring schemes which offer great value to the students who are mentored and also the students who do the mentoring. You can't beat explaining something to someone else, to help you understand it yourself.

Classes

I'm using the word 'class' to represent any scheduled encounter between a group of students and a member, or members, of staff. Classes can be set up in all sorts of ways and the words for them ('lecture', 'tutorial' etc.) are used differently in different situations. In some departments 'lecture' is used as a kind of default; it simply means a class, typically with quite a lot of students, which takes place at a fixed point in the timetable each week. This may be nothing like the traditional image of a 'lecture' in which the students sit listening and trying to take notes, while the lecturer speaks, perhaps pacing up and down at the front, staring at the ceiling for inspiration. 'Tutorial' is also used in lots of different ways. In detective dramas set in Oxford it describes a small group of students sitting with a lecturer (perhaps wearing a gown) in a room with a fireplace. But 'tutorial' is also used to describe quite a large class where students are working on examples and seeking help when they need it from staff.

In engineering and science (for example) there are lab classes, in architecture there are studio classes, in planning there are seminars. There are all sorts of ways of structuring support for project work and group work. There are 'workshops', some of which even take place in an actual workshop. There are practical classes, inside, outside, off campus. All these formats aim to provide a learning opportunity. They do this in lots of different ways, for different group sizes. While your lecture may be for 150 students together, your practical work may be with a group of four.

We're going to 'go behind the scenes' in Chapter 12 and look at why lecturers make the choices they do about how to deliver their courses. But just for now the aim is to help you prepare to get the most out of classes, whatever format they have.

You enhance your learning by what you do before, during and after each class. 'Before', 'during' and 'after' equate to *prepare*, *engage* and *follow up*, though really *engage* is the key word at all stages. *Preparing* is to do with engaging with material that has been provided in advance. *Following up* is really about study and practice, which we will cover next. Certainly don't focus only on 'during'. After my own degree studies I reflected that I had never managed to concentrate for the whole of a class – not once. I blamed my lecturers at the time, but I'm not so sure now (though lecturers in those days were much more boring than they are now).

Classes can be motivating, and some are inspiring. At the very least it is a human face to the subject and provides a spur to learning. Your only lasting record of what happened, of what you felt you learned at the time, is the material you take away. This could be printed material that you may have developed or annotated in some way, or your own separate notes. In my table above, this material is derived from an *experience* but it becomes a *source*, so it has a particularly important role in the learning process. That is the important thing about your notes and that is why you should treat them carefully. Forgive me for pointing out the obvious, but make your material and notes as neat as possible so you can follow them later. Put a date on everything in case you lose the sequence. Keep all your notes in a file that allows you to find things quickly. All this might be electronic, which may make the structure easier to control, but the organisation of the material is still important.

I know this is going to sound old-fashioned but I want to encourage you to take notes in classes, or at least annotate or personalise the printed material you are given in some way. It's tempting to think 'well, it's all here anyway, why should I bother to write stuff myself?' Well yes, the information is all there anyway, of course it is, it's also in books, on websites etc., but it's not the information that matters, it's your learning. And without notes you will have no record of your learning.

Attendance is important, though there is an interesting debate about whether it can be proved that attendance enhances performance, as considered later, in 12.1. It is true that for most courses there is a great deal of back-up material on the VLE (more below). But, unless you are studying by distance learning, you have chosen a mode of delivery which provides a face-to-face experience, and the more you engage with it the more you will benefit. Everything can't be on the VLE anyway. Something happens in a room full of people that doesn't happen in material on a website. You are bound to miss something. As a deadline or exam approaches I get emails from students who missed classes. I try to help, but there's no getting round the fact that they did miss something. Lecturers are familiar with the email from a student that more or less says 'I missed the classes and it's your problem'. But actually it isn't.

Study and practice

'Practice' might seem an appropriate word for studying a musical instrument, but for built environment or engineering? Well I think a lot of study in these vocational disciplines is 'practice'. It is a case of mastering techniques, typically of design, or quantitative analysis, or evaluating information, through studying cases or problems. So by studying, you're becoming good at doing something.

Studying in all subjects involves a lot of prioritising and planning. Professional people need to be good at this, so prioritising and planning in relation to study is a good place to start developing your professional skills.

Feedback on assessment

Forms of assessment are covered in 3.2. Feedback should, in theory, be the means by which you learn from assessment. From an engineering point of view the word 'feedback' itself is probably inappropriate. In an engineering system, feedback is used to adjust the input to a process. But academic feedback can't be sent back in time to adjust our work on a task; it can only be sent forward to adjust our work on the next task. So perhaps it should be called 'feed-forward'. The point is: that's what feedback is for – to allow you to improve your subsequent work.

When you access feedback on a piece of work you will probably be wanting to know the answer to three questions: 1. What mark have I got? 2. Did the marker think it was good? 3. What can I learn for next time? Question 1 is usually most important at the time and that's understandable – we're all victims of the system. Some staff think the mark is the only thing students are interested in and they point to the fact that whereas the marks are always accessed, the feedback is not (online, or by physical collection of the annotated work). The problem is that question 3 is nearly always masked by question 1 and maybe question 2. But of course in the grand scheme of things question 3 is by far the most important.

VLE and printed material provided

There are several levels of material:

1 The support material provided for specific modules on paper or via the VLE – this is clearly 'priority material', identified by your lecturer as particularly relevant.

2 General material about the course as a whole – this seems more boring, and is likely to be put to one side after induction or available in a part of the VLE you don't visit often. But read it – it will often contain some essential information that is not easy to find elsewhere. Make sure you know where to locate it for future reference.

3 Procedures, regulations – this seems even more boring. Best to know where it is – you may need it for something important.

Internet, books, documents

This material is discussed in Chapter 5. Reading may not be a favourite activity of built environment or engineering students, who tend to be attracted more to the technical, visual, creative or practical aspects of their subject. So how can you read effectively and efficiently? See if you find any of the following useful.

1 Don't worry if you're not the sort of person who likes to read something from beginning to end. Some people just can't – won't – don't engage with material in that way. Skim, skip, go back, go forward, if that's how you like to read. Read *your* way.

2 Point 1 applies to reading one source, or to several. If you like jumping from one source to another, don't feel there's anything wrong with that.

3 Don't confuse reading with memorising. You must make some notes to help you take in the material. If the copy is yours and you're planning to keep it, annotate it. (But of course it's not acceptable to make notes in library books or books borrowed from other people!)

4 If you're reading and getting nothing from it, stop. Reconsider what you're doing.

5 Don't even think about your reading speed. It simply doesn't matter. Read *effectively*.

6 You need to accept that some reading will be needed to find out if a source is relevant or not.

7 Obviously – as much as possible – concentrate on sources that are useful.

8 Use all the advice you can get, from lecturers, from fellow students, from anyone else, about sources that are likely to be most useful.

3.2 How do we show that we have learned something?

We're talking here about assessment. This is an area of study skills where the emphasis is on performance rather than learning, though there is still plenty of learning going on. Also, although professional life provides its own forms of assessment, really this section is more about study skills than professional skills.

Submitted work

It would be impossible to cover every form of submitted work here. This is more general advice about approaching the task and ensuring its completion on time.

Planning has to happen backwards. You must start with the deadline and work back to identify a reasonable time to give yourself. The other aspect of this, as we discussed in Chapter 2, is to establish priorities (with other pieces of work, other study activities, and everything else).

An important stage is to examine the task; try to be clear in your mind about what is required. You may need to go back to your lecturer with queries. Your fellow students may help too, but be careful that you're not just sharing misunderstandings. If you have not understood the task correctly it could affect your mark or grade.

> *Follow instructions; interrogate the brief; ask as many questions as possible, especially about assignments.* 🖓 **STUDENT**

If the work is a written piece, one thing you need to be clear about is the *structure* of what's expected. Some types of reports have very standardised structures and you must find out about this before you get too far with writing. The structure of a report is always explicit: your readers should be able to see what the structure is before they start actually reading, by looking at the subheadings and (depending on length) the contents page. There's much more about this in Chapter 13.

Some built environment subjects, for example planning, housing and to some extent architecture, have elements of submitted work in the form of **essays**. Essays must have a strong structure, but it is implicit: you need to read an essay to appreciate its structure. This needs just as much care in planning, but a different approach to writing.

As you begin to work you need to collect your thoughts. Develop ideas in your head and on paper. Make notes about content in relation to structure. Don't start trying to produce something until you know what you plan to do. The next chapter (Chapter 4) is all about writing. It includes basic material about punctuation etc., and also guidance on how to get started and how to develop ideas.

Revising for tests and exams

In some courses, tests and exams form a significant component of overall assessment. Nobody likes taking exams, but anyone can have success if they think about how to achieve their best. First we'll look at revising for exams and then at actually taking them. I am deliberately keeping this short. There is plenty of advice out there if you want more, from lecturers, from books on generic study skills, from study-support websites etc.

Revising really means preparing, and you should start preparing for a test or exam at the start of the module. As the exam approaches your study activities become more focused on the exam itself, but there should be no sudden change: preparation should be continuous. As part of the preparation, there are some things you should do:

1 Sort out all your material. You may have been doing this week by week anyway; of course this is the best way. But if you've got behind with your 'filing', sort things out into a logical sequence and ensure everything is easy to find.
2 Actively test your understanding of topics or calculation techniques. Don't just 'go over' material.
3 Look at previous papers to understand the format and scope of the questions. Find out if there are going to be any changes from previous years.
4 Attend 'revision sessions' and listen to guidance from your lecturer.
5 As the dates for exams approach and there is more time for revision, plan your days so that you cover all subjects but also have time to rest.

Taking tests and exams

It is the Olympic 100 metres track final. The athletes are standing in their lanes. Four years of preparation for 10 seconds of performance. They are intensely focused and yet outwardly calm. As their name is announced they smile, wave at the camera, then go back into their zone of preparedness.

Watch sports people at the start of important events – this is how you should be just before an exam. You should be calm, well prepared, quiet, focused. Don't engage with other students as you wait to go in the exam room, don't get involved in over-excited conversations, 'have you revised xxx? what's the formula for yyy?' Don't tire yourself before the exam with last-minute revision. Your preparation should have started months ago.

In the exam, plan ruthlessly. Read the whole paper carefully, but then attempt the questions that seem easiest to you first. Think about what each question is really asking. Balance your efforts by looking at the mark allocation. If you've run out of time for a question, stop, and only go back if there's time later. If you finish early, don't leave until you are sure you can't do more and have checked everything carefully.

4
Writing

4.1 Basic writing skills

Many students – not only those involved in built environment and engineering – lack confidence in writing. You may be one of them. So here are some aids that you may find useful.

Words

A lot of checking gets done for us – by our computers or phones. But we still need to know what's right without these aids, for example when:

- the spell checker corrects the spelling but gives you the wrong word
- you are summarising a group's discussion on a flip chart
- you are writing a quick note to someone
- you are checking something on paper
- you are making a sign in a hurry

Spellings: here are some to watch (correct spelling given here)		
analyse analysis anomaly apparent beginning develop diagram exaggerate forty fourteen	gauge government grammar height install instalment liaise maintenance necessarily ninety	occasionally omitted parallel proceedings preceding professional simultaneous separate successfully unnecessary

Test yourself

Some of the spellings below are correct and some are incorrect. Spot the words that are misspelled and provide the correct spelling.

accommadation
arguement
disasterous
enviroment
fulfil
independent
ocurred
procedure
superceeded
wier

Answers on p177

Parts of speech		
Noun	thing / person	*theory building friend*
Verb	doing / being	I *walk* to the campus. This *is* pointless.
Pronoun	goes in place of a noun	*you me they*
Adjective	describes a noun	*big* changes This activity is *pointless*.
Adverb	describes a verb or an adjective	working *efficiently* *completely* pointless
Conjunction	links parts of a sentence	*and but*
Preposition	indicates position: answers the question 'where?'	*in on to*
Article	introduces a noun	*the a*

Words to watch		
affect / effect	*affect* is a verb, *effect* is usually a noun	Thickness *affects* thermal resistance. This has a significant *effect*.

verbal / oral	*verbal* means *in words*, *oral* means *in speech*	A lot of people think *verbal* means spoken and not written down, but strictly it doesn't (a letter is a form of verbal communication).
principal / principle	*principal* is usually an adjective meaning *major*, a *principle* is a *rule* or *natural law*	*principal* stress Archimedes' *principle*
relative / relevant	*relative* is often used when *relevant* (*of particular interest*) is the right word	*relative* density Let's consider the *relevant* information.
complement / compliment	a *complement* fits well or makes something complete, a *compliment* is polite praise (both can be verbs or nouns)	She *complements* the team perfectly. I must *compliment* you on the quality of your presentation.

I saw this sign on the road in the college where I work. At first I thought the bottom line meant 'look, just deal with it; you may not be able to get where you want but we've done our best', but then I realised they had used the wrong word. They could even have checked it on a spell-checker before writing the sign.

Apostrophe

This marks either a missing letter (*doesn't*) or a possessive (*Sam's idea*). For plurals the apostrophe comes after the *s* (*engineers' talents*). Be careful about *it's* and *its*. *It's* is short for *it is*; the possessive *its* doesn't have an apostrophe (*It's the best option and its benefits are obvious*).

What's gone wrong here? Did they forget to put a full stop after GRANNIES?

Punctuation

Punctuation Item	Name	Where to use (examples below)
.	full stop	end of a sentence
,	comma	pause in a sentence
;	semi-colon	break in a sentence
:	colon	introduces a phrase or list
-	hyphen	joins words together; useful for clarity
?	question mark	must have at the end of a question
'...'	inverted commas	for quotes

Some punctuation examples

We must find a solution, one that satisfies everyone.	*The comma helps with clarity but the two parts of this sentence could not be separate sentences.*
It is not a new idea; it has been around for years.	*The two parts here could be separate but are closely related.*
There is one rule when writing: be clear.	*A colon is best here, not a semi-colon.*
There will be a series of two-day tests.	*The hyphen joins 'two' and 'day' to make one word.*

Sentence formation and punctuation

Not good	Corrected
This report discusses the topic of flood management in the UK. Considering hard and soft engineering techniques in use today. As well as possible future developments.	This report discusses the topic of flood management in the UK, considering hard and soft engineering techniques in use today, as well as possible future developments.
Flooding across the UK has become more severe in the last few decades, this is due in large part to climate change over 5.4 million properties are at risk from flooding. And 2.8 million properties are at risk from surface-water flooding alone.	Flooding across the UK has become more severe in the last few decades. This is due in large part to climate change. Over 5.4 million properties are at risk from flooding, and 2.8 million properties are at risk from surface-water flooding alone.
More and more sustainable technologies are being produced from photovoltaic cells to ground source heat pumps.	More and more sustainable technologies are being produced, from photovoltaic cells to ground-source heat pumps.

Test yourself

Spot the errors, and correct by using better punctuation.

1 What is the explanation.
2 These are examples of poor punctuation they show what can go wrong.
3 Bad punctuation causes two main problems; ambiguity and poor readability.

Answers on p177

Paragraph

A paragraph is a group of sentences that fit together. A new paragraph introduces a slightly new theme. Paragraphs are separated by a vertical space on the page (either using paragraph format or by pressing the Enter key twice). In books, paragraphs tend to be separated by indenting the start of the new paragraph, but that is not common in reports.

4.2 Getting started

I'm not going to set down rules for you here – just the opposite. The important thing is to find something that works for you. And it really is important. If you can become a fluent writer when you weren't before, that could make a huge difference to your ability to explain, argue or convince. If you think you 'hate writing' you might be able to change that; think how useful that would be.

Sometimes when we approach a writing task we have a pretty good idea of what we want to say, but we are not sure exactly how to word it. At other times we just don't know what we want to say.

If you don't know what to say it is a bad idea to force yourself to write, only to end up just staring at a blank screen or piece of paper. You should try to let your thoughts run free and capture your ideas as notes (more about this shortly).

If you do know basically what you want to say, don't inhibit yourself by feeling you should make notes or spider diagrams, or anything else you remember being told you should do. Sometimes it is best just to write. But don't feel you are writing some kind of finished article. Many years ago I might have thought of a form of words for something, and then thought 'that's not good enough – it's going in a report!'; now I just write and think 'I've no idea if this will end up in the report, I think I'll just write it anyway'. In fact it's probably better not to think about how it will end up: as a report, a dissertation, whatever. Just think of a writing task as an organic entity that will grow, perhaps in unexpected ways, and may be pruned, or taken apart and put back together. Or maybe think of writing as slapping down some clay; the moulding process (to form a pot, or a dissertation) can come later. Of course you have to know what you want to achieve to some extent before you can do the slapping down, otherwise you wouldn't know how much clay, or what type of clay, to use. But if you are getting anxious and stuck trying to mould the words of your essay or dissertation, even though you feel you know basically what you want to say, just slap down a section or two, very rough, and you'll feel much better.

But back to what to do if you don't know what you want to say. A stage in writing that can be very beneficial is to **open up** the topic. This involves bringing in everything that could be relevant, all your ideas, all possible sources. This may be cumulative and rather random, and does mean you'll probably need to start with notes of some form.

This opening up is more effective if you're not trying to find the exact words at the same time. I know that's not much help if you are already up against a tight deadline. But then maybe a lot of bad writing is really just bad time management. Just using precious time to 'open up' may seem like a luxury you can't afford. But I learnt about opening up very vividly once and I've always remembered the lesson.

I had decided to apply for a new role at work. I had been working on the application, filling in bits of the form (which had a very complicated format), but I wasn't really getting anywhere. I had tried to think of relevant material to add, but I'd just dried up. I found myself wondering if I had much to offer after all; there didn't seem to be anything to say.

It was the start of my summer holiday with my family. We'd booked a holiday in Greece in a beautiful place with a terrace overlooking the sea. I loved just sitting there in the shade letting my thoughts wander. I hadn't planned to work on the application while I was away, but I found myself thinking about it without any focus or pressure. Then I started having some ideas – sources I hadn't thought of before, all sorts of things I could include. Then even more ideas came – it was really opening up. I didn't have any proper paper to hand so I made notes on the back of a paper bag. Sitting there during the holiday I found myself adding to these ideas, opening up far beyond the words I had been painfully trying to force out before. It felt like inspiration, a transformation, but it was just the result of making space to explore ideas.

Unfortunately we can't book a holiday every time we have something to write. And, if we did, we'd probably have better things to do. But the principle is this: when you're not sure what you want to say, if you can find some space to think, if you open up instead of forcing words, you may get some great results.

Maybe the best way to look at it is this. If you've got an idea of what you want to say, and your inclination is just to go to the computer and write, then do it without trying to get it right first time. But if you don't have much idea of what to write, it is better to open up and make notes, and not to force yourself to write.

4.3   Getting closer

Even if you have made notes first, when the time comes to write, just write rough, don't try to get it perfect.

Once I was supervising a final-year project and the student brought me a draft chapter to read. He had been trying to write in a very formal style. This had obviously taken him ages but the writing was clumsy and distracting. I asked 'you don't normally write like this, do you; is this your posh style?' He nodded and said, 'well, it is my final-year project'.

So of course I talked him through the ideas you have probably just read: 'don't worry about how it will end up', 'slap it down', etc. I'm sure it took him no longer to write the whole of the rest of his report than it had to write that first chapter. In fact the results of rough writing are often just as good as fussy writing, even though that doesn't matter anyway – the job isn't finished in either case.

The final stage, let's call it '**moulding**', is the most important. This is when the final written product is created. It shouldn't feel like hard work; after all, you've done the writing. I recommend trying to create as much separation as possible from the writing stage. Do it somewhere else, and, if you can afford the time, do it after a bit of a break. This moulding process basically involves reading and revising. Even if you are doing it on a computer, try to 'sit back'. You will do this more than once, so even now don't think 'it's got to be perfect'. Just read, and amend as you see fit. It's only when you've read it and said 'yes, it's OK', that the process is complete. This process works especially well if you are right away from 'work' – sitting outside, or on a train journey perhaps.

4.4   Style

Documents and reports should be written in a simple, clear style. They must be professional and credible, which means they should be fairly formal, but not so formal that it stops them from being instantly understood by the reader. By the way, I need to emphasise that this book is written in an informal style. It is not the style I would use to write a report.

Common advice for students is that reports should 'not be written in the first person' and they 'should be written in the passive voice'. What's that all about?

First we need to consider basic sentence construction:

>   subject – verb – object

Here's a sentence:

>   'I measured the depth of water using a point gauge'.

The *subject* is 'I', the verb is 'measured' and the *object* is 'the depth of water'.

In any sentence, the subject stands out because it often comes first. But in this sentence the subject (the person who did the measuring) isn't important at all.

This sentence can be turned from the *active* voice to the *passive* voice by writing:

>   The depth of water was measured using a point gauge.

In fact I should strictly have included 'by me' in that sentence, but I have taken the opportunity to drop it because it's not needed.

'I' or 'we' is called *the first person* and it is normal in formal writing not to use it. The device for avoiding it is the passive voice.

But there is no need to write every sentence in the passive voice. Here are some entirely appropriate sentence constructions that are in the active voice.

>   This report presents an investigation of...

>   The aims are to...

>   Seawater can have a corrosive effect on...

>   The main findings of this investigation are...

To achieve an appropriate level of simple formality it is also a good idea not to use common contractions like 'doesn't', hasn't', etc. Just write these as 'does not', 'has not'.

Do not use colloquialisms like 'there's no way that ...', or 'What's that all about?'

There are examples of the appropriate style for a report in 13.3 and for a research report in 14.7.

Formal style in a report – Summary

Avoid the first person: 'I' or 'we'
Use the passive voice to avoid the first person
You don't have to write everything in the passive voice
Avoid contractions: 'doesn't', 'hasn't', etc.
Avoid colloquialisms such as 'there's no way that...'

5

Information sources

5.1 Sources

If you think too hard about it, an academic library can seem a scary place. Any book in it could be dense, impenetrable, maybe give you evenings of labour to battle through the first chapter. And yet there are hundreds of thousands of books. How much could anyone read? There's knowledge here, but how much could we turn into understanding? For us, what tiny percentage is really accessible?

The internet is scary in a different way. In fact in a lot of ways it's scarier than the library. At least in the library most of the material can be considered authoritative in some way. On the internet – or just basically 'out there' – there is so much information, produced by anyone, everyone really, that the amount is more or less inconceivable. I admit I don't know how to imagine it or think about it. And of course it's always growing and always changing. There's information, but how much could we identify as useful knowledge and then turn that into understanding?

Maybe it would be better to approach this the other way round. What do we really know about sources of information, and how can we have confidence that they are valid?

Books
We may have confidence in the content of books because they have been recommended to us. A book has to be accepted by a publisher and this is usually done through an expert review system. But publishers are commercial organisations, not academic standard-setters.

Journals
Academic journals usually have quite a strict reviewing system, so a journal paper has normally been checked by subject experts before it is published.

Academic conferences

These papers are also reviewed by academics, though for most (but not all) conferences the reviewing process is less strict than for journals.

Publications by national organisations

It's impossible to generalise here. If you know about the organisation, then you know how much confidence you can place in the content of its publications. But if you don't, you could be looking at a publication written by a biased pressure group, or a commercially-motivated trade organisation.

Theses

Theses, especially those written as part of a PhD research degree, can be very useful. No one has access to theses of candidates who failed, but even in a successful thesis everything may not be correct, so some care is needed.

Websites

Whose website? Government department, respected national body – fine. The weird bloke next door who collects steering wheels – more problematic. Some open web resources to which anyone can contribute can be very useful, but caution is obviously needed about quality and reliability of information.

5.2   Searching

I've learnt that above the age of about eleven or twelve, the younger you are, the better you are at searching for something online. So I'm not going to say much here. The one thing I would say however is that when searching for academic material it is hugely helpful to know about search engines that are designed for that purpose. You are almost certain to have 'information skills' sessions with a librarian or information-skills adviser and you will be told some incredibly useful stuff. I find many students attend these sessions but then, if they don't start using the techniques straight after, they just revert to the information-retrieval methods they use for the other parts of their lives. The problem is that you can get quite a lot of good stuff using everyday methods, but not everything that is important. My project students say they can't find something I have mentioned and I say 'why not use the searching techniques you were taught?' But they usually give me a bit of a blank look. You can always go back to the library or information resources centre and ask for advice again. Or look on the information resources website.

5.3　Choosing

There's too much out there – that's always the problem. You must decide:

Is it relevant?

This is a decision for you, but don't fall into the trap of including something just because you've found it. The sources must suit the purpose, not the other way round.

Is it trustworthy?

In other words is it basically accurate, correct, well-balanced, or is it biased or simply wrong? We've covered the sort of thing you might consider in our discussion of sources above.

Is it sufficiently up-to-date?

In a sense this is where online information might have an advantage over printed sources. It's just a matter of finding the year of publication and deciding if that is recent enough for your purpose.

5.4　Referencing

In what you write you must acknowledge all your sources. This is so that:

1　No one can accuse you of passing off someone else's work as your own (plagiarism – covered in detail at the end of this chapter).
2　You can show off about the range of sources you have used (and how good you are at referencing).
3　You don't have to write something that somebody else has already written about.
4　Someone wanting to find out more about a particular topic will know where to look by following your reference.
5　The person who wrote the work is given due credit for it.

The most common system of referencing in work at university, by far, is the one that is generally called the 'Harvard system'.

In the Harvard system you 'cite' a publication in your text by giving the author's surname and the year of publication. You either put the name and year together in brackets, or if the name appears naturally in the sentence you put the year in brackets straight after. When the reader sees this they know they can find the details of the publication in your list of references (usually at the end of your piece).

Examples of citing (Harvard system)

This has also been demonstrated at lower temperatures (Atkinson, 2014).

This was also found by Atkinson (2014) working at lower temperatures.

In the list of references some essential information about the publication is provided so readers can find it themselves.

Book	Author(s) [surname and initials], year [in brackets], book title [in italics], edition [if needed], publisher.
Contribution in a book	Author(s) of contribution [surname and initials], year [in brackets], title of contribution, 'In:', editor(s) of book, title of book [in italics], edition [if needed], publisher, page numbers.
Journal paper	Author(s) [surname and initials], year [in brackets], title of paper, name of journal [in italics], volume number [bold], issue number [in brackets], page numbers.
Conference paper	Author(s) [surname and initials], year [in brackets], title of paper, 'In:', title of conference proceedings [in italics], volume number [bold], location of conference, page numbers.
Thesis	Author [surname and initials], year [in brackets], title of thesis [in italics], degree for which submitted, institution.
Report	Author(s) [surname and initials], year [in brackets], title of report [in italics], report number, organisation.
Web material	Author(s) [surname and initials], year [in brackets], title of material [in italics], web address, date you accessed the site [in square brackets].

Using the Harvard system the references are listed in alphabetical order of the first author's surname regardless of the type of source or the order of the citations in the text.

Examples of references (Harvard system)

Atkinson B.W. (2014) Impact of activity levels on thermal comfort in canvas structures. *Journal of Building Materials,* **23** (4), 125–131.

Dandini G.G., Parmison J.B. and French K.T. (2016) *Principles of Control,* 4th edition, London: Palgrave.

Edgar F.N. (2015) Finite element modelling of air flow in filing cabinets. In: Cho J.Q. ed., *Developments in Finite Element Modelling,* Spon Press, 213–245.

Grimm A.S. (2017) *Allowing for climate change in green roof design.* PhD thesis, University of North Devon.

Ignatious V. and Adenle R.W. (2014) Teamwork in design projects, an old problem revisited. In: *Proceedings of the 5th International Conference on Design Education,* **2,** Coventry, 89–93.

SUDS Society (2015) *Photo gallery: swales,* www.sudssoc.org.uk [accessed May 2016].

Williams I. and Jones D. (2017) *The state of precision manufacturing in Wales,* Report W234. Welsh Council for Engineering, Cardiff.

In the list above note that the references are not numbered – the order is alphabetical: numbers are not used or needed.

It's easy to get confused about which bit gets put in italics. The rule is that it is the name of the volume that you would take off the library shelf (or equivalent). So it's the title of the book, journal or report, but it is not the title of the journal paper or conference paper.

Where there are more than two authors (as in *Principles of Control* above), the citation can be (Dandini *et al.*, 2016) where *et al.* is short for *et alii* which means 'and others' in Latin. All authors are given in the list of references.

If you are *quoting* from a source, you should specify the page number when you cite:

> At lower temperatures 'activity levels are particularly significant'
> (Atkinson, 2014: p126).

Where the author is an organisation, the name of the organisation is used in place of a surname:

> (SUDS Society, 2015)

Where there are two references by the same person in the same year, the citations become:

> (Edgar, 2015a)
> (Edgar, 2015b)

Then the 2015a and 2015b appear in the list of references so the reader knows which is which.

Images like diagrams and photos, and tables, that are not your own, must be referenced. The citation can be placed either in the caption for the image or table, or in the text at the point where reference is made to the image. The source is detailed in the usual way in the list of references. It is usually better to create your own images anyway (and definitely better from the perspective of getting a good mark), but if you use someone else's image you must provide the citation and reference, otherwise it is plagiarism (as discussed in 5.5).

Citing an image

Caption:

Figure 8.2 A typical swale (SUDS Society, 2015)

Text:

Typical use of a swale is shown in Figure 8.2 (SUDS Society, 2015).

You will be amazed by the sheer fussiness of lecturers when it comes to references. The rules are simple, you just have to apply them carefully. But I think students are caught off-guard by the emphatic insistence, by people who mark their work, that such little things really matter. But, hey, if only everything was that simple. Apply the rules, and get that tick in the inevitable box on the marking sheet that says 'Harvard referencing correct'.

Example of using a source

As part of a project, you have designed the stormwater drainage system for an industrial estate. You are writing the project report and you want to point out that your design is part of a 'separate system' where stormwater and wastewater are drained separately, as distinct from the more old-fashioned combined system. You have read the following in a book called *Draining Cities* by Colin Sewellson (published by Diameter Publishing in 2015):

"In the combined system, wastewater and stormwater are carried together in one pipe system. In the UK, most of the older sewer systems (built before 1945) are combined, 70% of the total by sewer length. During rainfall some of the combined flow may need to be 'overflowed' to a river or stream at a structure called a combined sewer overflow, or CSO, and because the overflowed water is a mixture of wastewater and stormwater, some pollution is inevitable. It is mainly for this reason that more recent systems are separate, with wastewater and stormwater in separate pipes. These are generally the systems built (in the UK) since 1945, 30% of total length. They avoid the pollution of CSOs. There is extra cost compared with a combined system, but not twice as much because the pipes are usually constructed in parallel within the same trench."

This is not a major aspect of your work but you feel it is worth a mention. Here are three ways you could use and reference this material in your report.

1   You could simply refer to the fact that the system you have designed is *separate* as opposed to being *combined*, and provide a reference to the book for anyone who wants to find out more.

   The drainage system described here is a 'separate system' (with the stormwater being carried in a separate pipe system from the wastewater) as opposed to the more old-fashioned 'combined system' (Sewellson, 2015).

2   You could give a bit more information by summarising or paraphrasing what you read in the book. Now the reference is important not only to help your readers find out more but to formally acknowledge your source.

The drainage system described here is a 'separate system' (with the stormwater being carried in a separate pipe system from the wastewater) as opposed to the more old-fashioned 'combined system'. In fact 70%, by length, of UK sewers are combined, but they can be a cause of pollution because during rainfall there may be overflow to the river or stream. Separate systems have been built since 1945 to avoid this problem. They are more expensive to construct than combined systems but the additional cost is limited because the two pipes are generally constructed together (Sewellson, 2015).

3   You could also include a quotation from the book. This can provide a sense of 'authority', though you should only quote if there is a good reason. If you quote too much you could get a lower mark simply because less of the work is your own. The citation following the quote includes the specific page number, though it is not just the quote that needs to be referenced.

The drainage system described here is a 'separate system' (with the stormwater being carried in a separate pipe system from the wastewater) as opposed to the more old-fashioned 'combined system'. 70%, by length, of UK sewers are combined, but they can be a cause of pollution. 'During rainfall some of the combined flow may need to be "overflowed" to a river or stream at a structure called a combined sewer overflow, or CSO, and because the overflowed water is a mixture of wastewater and stormwater, some pollution is inevitable' (Sewellson, 2015: p78). Separate systems have been built since 1945 to avoid this problem. They are more expensive to construct than combined systems but the additional cost is limited because the two pipes are generally constructed together (Sewellson, 2015).

In all three cases, the list of references will include:
Sewellson C.J. (2015) *Draining Cities*, Diameter Publishing.

Test yourself

Citations and references

You are writing a piece on referencing in the context of student writing. In one of your paragraphs you want to discuss students' attitudes to referencing. Your research has identified three good sources.

In a book called *Students Can Write* by Edgar Huntsman, published in 2013 by Apple Press, there is a good quote on page 13:

It seems clear that students understand that correct referencing is important (not least in terms of assessment) but they cannot help feeling it is trivial.

You have found some stats in a report published by the Higher Education Academy that seem relevant. Report BB21, 'Academic writing: a fairground ride', says that in a survey of 154 students at a university in the Midlands, 61% felt that correct referencing improved quality of writing, and 85% felt that correct referencing improved assessment of writing. The authors of the report are identified as Andrea Wiltshire and Rik Ghent, and the report is dated 2012.

The website of the organisation 'Student thirst', www.studentthirst.org.uk, which you visited on 12 December 2014, has some quotes from a survey on referencing, most of which consider good referencing to be necessary but boring. One is:

> I honestly think lecturers care more about referencing than ideas.

This content is on a section of the website with the title 'About Referencing'. At the bottom of the webpage it says 'Last updated May 2014'. No individuals are identified as authors.

Write the paragraph including *citations*.
Write out the *references* **that will go in your list of references.**

Suggested answer on p177

5.5   Plagiarism

Plagiarism is massive. I mean three things by that. Firstly it is a big deal for universities. Most have very comprehensive policies and procedures for dealing with plagiarism: discouraging it, detecting it, and penalising those who commit it. Secondly there is so much opportunity and temptation around, with so much text available and such easy ways of copying it. Thirdly it can be massively serious for people who are identified as having plagiarised. Fortunately (as we will see) there are ways of finding out if you have plagiarised unintentionally.

Plagiarism is a form of cheating, and it usually involves using someone else's words in a written submission without making it clear that they are not yours. Typically someone finds a chunk of text on a website that describes exactly what they need to write about, and, perhaps because they are running out of time, simply copy and paste the text into their own work – a few clicks instead of hours of work. They submit the assignment with the implication (and usually these days an actual signature to affirm) that it is their own work.

When students are identified as having plagiarised, and detection is easy with software, there can be heavy penalties. In very serious cases this can be: fail, and don't come back. I sense that some students are surprised by this and feel that some subtle distinctions are being made. If the copied chunk of text had been placed in inverted commas, making it clear it was a quote, and this was followed by a citation, with full details of the website in the list of references, this would not be plagiarism. It still might not have got a good mark because a large chunk was just a quotation, but there would be no other problem. Citing and referencing are crucially important when you use other people's material.

There are other related forms of cheating. Collusion is when you work with other students on a submission that was supposed to be completely individual, or copy another student's work, or (ouch!) allow another student to copy your work. And of course it is cheating to use someone else's work in its entirety: a student from another year, or someone you've paid. Enough – I'm getting depressed. Listen carefully when they tell you about all this, and don't do it!

The software that detects plagiarism can be a very useful tool for you to use yourself. You may be required to do this at an early stage before final submission – it's a very good idea. It will alert you to any content that could be perceived (rightly or wrongly) as plagiarism, and it will tell you if you have plagiarised by accident. If you find this to be the case then you would be wise to get some expert advice on appropriate use and attribution of sources, rather than trying to reduce the similarity percentage by manipulating the copied text.

From every perspective: be careful about plagiarism.

6

Number and data presentation

All the built-environment and engineering disciplines have some call for skill with numbers and calculations. For example, architects, architectural technologists and certainly building surveyors must have an understanding of the structural and scientific behaviour of buildings; quantity surveyors and project managers must have a highly quantitative grasp of the management of cost and time; construction managers must make accurate predictions about materials or temporary works. Perhaps the need for numeracy skills is most obvious for engineers, but this chapter is definitely not just for them.

Though they don't have to be mathematical geniuses, engineers (and many other built environment professionals) must be comfortable with quantifying problems and solutions. A lecturer who taught me, and who I didn't like very much, used to say 'engineers are scientists who've got to get their sums right'. I suspected at the time that this was his excuse for not going through students' calculations on exam scripts if the answer was wrong. (Don't worry – there's no one like that left now.) But he did have a point about real engineers. A well-thought-out solution is no good if a miscalculation has made the slab too weak or the fuel mixture too explosive, or whatever. I once worked on a site where a tunnel completely missed the shaft that it should have connected with, because the setting-out engineer miscalculated the bearing by one degree. At least no one was harmed (apart from the engineer's self-esteem and immediate career prospects).

A lot of quantification in engineering and built environment is fairly straightforward; a lot of it is just estimating. It may not involve fancy maths, but you've got to be sure of your answer. Professionals must know how to go about checking that their estimates are right, and for more complex calculations must make sure there's a system in which everything is checked.

6.1 Numbers and symbols

Estimating

I was training to be a reviewer for a professional body (someone who interviews candidates to determine whether they should become professionally qualified engineers). I was shadowing a review panel of two senior engineers to see how they did it. It's scary for the candidates because it is so important for their careers and because they can be asked more or less anything. This candidate had designed part of a wastewater treatment works. The lead reviewer asked about dilution of the effluent when it was released into the natural stream. The candidate didn't have details about this, which didn't seem to be a problem in itself. The reviewer, to give a bit of a prompt, said 'well, what's the typical flow-rate in the stream – just roughly?' The candidate looked a bit shocked: he'd obviously been taken by surprise by this. I felt sorry for him – all that preparation, just to be caught out by a simple question. 'Just a rough figure', repeated the reviewer. The guy obviously needed to come up with a figure somehow. I found myself wondering what *I* would say. The stream looked about 2 m wide from the plan. How deep? Maybe half a metre average. Flow-rate is area times mean velocity, so how fast? 1 m/s is a moderate velocity. So $2 \times 0.5 \times 1$, not a difficult sum: $1 \text{ m}^3/\text{s}$. At that moment the candidate said 'about $1 \text{ m}^3/\text{s}$ I would say'. Relief all round, a few more questions, and he passed – an estimate that he had to get right in a pressured situation.

Of course there's pressure too in the design office, in the board room, in the workshop, on site, in the lab, on the shop floor. Professional people are making estimates that have to be right.

Test yourself

No calculators!

1 The density of water is 1000 kg/m^3. A 20 mm depth of rain is standing on a flat roof 10 m × 5 m in plan. What is the total mass of water (in kg)?

2 The unit of electrical energy used in electricity bills is the kW h (one kW for one hour). A 100 W lamp is left on by mistake for 100 hours. If electricity is 15 p per kW h, how much does that add to the electricity bill?

3 Five people live in a house. The water bill for a three-month period shows they have used 90 m^3 of water in total. How does their water consumption compare with the UK average of around $0.15 \text{ m}^3/\text{person/day}$?

Answers on p178

Accuracy

In my story above about estimating the flow in the stream, it would have been strange if the candidate had said '1.13 m³/s'. Simply by saying 1 m³/s he was implying that the number was an estimate. Strictly this implies a number greater than 0.5 and less than 1.5. This range might be too large for our example, though the number 1.0 would be inappropriate because that implies too small a range: 0.95 < value < 1.05. Accuracy is conveyed in this way. My students sometimes write '0.457 (to 3 d.p.)', but the bit in brackets is completely unnecessary. And of course it's not the number of decimal places that matter as much as the number of significant figures. Students with a poor understanding of this aspect of maths, who have calculated a number to be say, 0.00338, sometimes write it as 0.003 – after all, they might say, it's a very small number anyway, and is given with an accuracy to 3 decimal places! But the obvious point is that this will have a significant effect on the accuracy of any further calculations. Yet writing 0.00338167353 would be inappropriate in many situations because it implies an accuracy that does not exist. This can be difficult if the answer to a calculation is something like 227 864. You could write 228 000, but some people would say 'but that's not the answer'. Your maths teacher would tell you to write 2.28×10^5, but that might not feel very natural.

Test yourself

Calculator allowed

1 Round 12.849 to three significant figures.
2 Measure the long side of an A4 sheet in mm. If you were dividing an A4 sheet along its longer side into 7 strips of equal width, how wide would each strip be? If you were asking someone to mark out using a ruler and pencil where each strip should be cut, precisely what width would you specify?
3 How many seconds are there in September?
4 Is it OK to say that since 0.024 is such a small number we might as well write it as 0.02? If this number was someone's pay (in p) per second (averaged over the whole time, i.e. when they were working and when they weren't), how much difference would it make to how much they earned in September?

Answers on p178

Units

Units matter, of course. If the answer is 2.34 m then it is definitely not 2.34 mm. Specifying the correct units for any particular parameter is beyond the scope of this book. We will deal with the basic conventions about units generally.

Most but not all countries use SI (*Système International*) units. There are seven base units which have standard abbreviations:

metre (length)	m
kilogram (mass)	kg
second (time)	s
kelvin (temperature)	K
ampere (electric current)	A
candela (luminous intensity)	cd
mole (amount of substance)	mol

There are 22 'named units' derived from these base units, each with a standard abbreviation:

Radian	rad
Steradian	sr
Hertz	Hz
Newton	N
Pascal	P
Joule	J
Watt	W
Coulomb	C
Volt	V
Farad	F
Ohm	Ω
Siemens	S
Weber	Wb
Tesla	T
Henry	H
degree Celsius	°C
Lumen	lm
Lux	lx
Becquerel	Bq
Gray	Gy
Sievert	Sv
Katal	Kat

Although many of the units are famous scientists' and engineers' names, the name of the unit is not written with a capital letter. Yet in most cases the abbreviation *is* written with a capital letter. So the unit of force named after Sir Isaac Newton is the 'newton', abbreviated as 'N'.

Other common units and abbreviations are:

angle:
degree   °
minute   '
second   "

area/mass/volume:
hectare   ha
tonne   t
litre   l

time:
day   d
hour   h
minute   min

Since the abbreviations are standard it is best to stick to them, and sometimes downright ambiguous to deviate from them. Capital 'M' doesn't mean metres, for example, only 'm' does. (Make sure your word-processing software is not capitalising units for you when you don't want it to.) There are no plurals. 'kgs' doesn't mean kilograms. Units should not be italicised. That would cause confusion between units and symbols (as we will see shortly).

A space should be left between the number and the unit:

25 m
30.6 N
3.24 s

We all know km stands for kilometre, 10^3 m, and mm stands for millimetre, 10^{-3} m. The common standard prefixes, with abbreviations, are:

kilo	K	10^3
mega	M	10^6
giga	G	10^9
milli	m	10^{-3}
micro	M	10^{-6}
nano	N	10^{-9}

You can see that the distinction between upper case and lower case is crucial here.

You'll be familiar with cm (centimetres) from school and from buying clothes, but this unit is not favoured in engineering or built environment since multiples of 10^3 or 10^{-3} are preferred.

Units are commonly combined (beyond the combination of base units to form named units). Velocity is length divided by time, in metres per second. Engineers tend to write this as m/s though scientists would prefer m s^{-1} or m·s^{-1}. While you are a student it would be best to follow the convention you are given.

All of the named units can be expressed in terms of the base units. The named units are usually more convenient, though sometimes the base unit form is common too. For example, the pascal is the unit for pressure, equivalent to newtons per metre squared. But it is quite common to express pressure in N/m^2 anyway.

Symbols

You will quickly pick up the symbols that are common in your discipline. It is important to use symbols consistently, and define them clearly if they are not standard.

When letters are used as symbols it is standard practice to italicise them.

Many mathematical symbols are Greek letters, so here is a full list for reference:

Letter	Capital	Name	Letter	Capital	Name
α	A	alpha	ν	N	nu
β	B	beta	ξ	Ξ	xi
γ	Γ	gamma	o	O	omicron
δ	Δ	delta	π	Π	pi
ε	E	epsilon	ρ	P	rho
ζ	Z	zeta	σ	Σ	sigma
η	H	eta	τ	T	tau
θ	Θ	theta	υ	Y	upsilon
ι	I	iota	φ	Φ	phi
κ	K	kappa	χ	X	chi
λ	∧	lambda	ψ	Ψ	psi
μ	M	mu	ω	Ω	omega

Example of equation using italicised symbols and Greek letters:

$$H = \frac{p}{\rho g} + \frac{u^2}{2g} + z$$

Setting out equations in a word-processed document is much easier using the 'equation editor' facility. The results of trying to create equations of any complexity using standard punctuation usually look terrible.

6.2 Presenting data

(x, y) graph

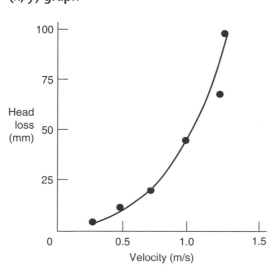

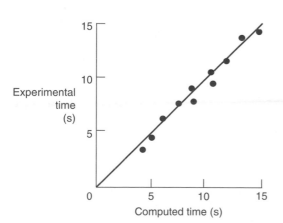

Of course you've been plotting graphs since you were at primary school. You can explore the relationship between simple sets of data by hand on a sheet of graph paper, or with more complex data, or for better presentation, using a spreadsheet. Typically there are points plotted from the (x, y) scales, and lines (straight or curved) that are used to make sense of the data. Always be clear about the relationship between the points and the lines. A line may be based on the data – a line of best fit for example, as on the upper graph. Alternatively, the line may represent a particular relationship against which the data is being compared. On the lower graph here, the line is a 45° line representing (for equal scales on the x and y axes) equal values of experimental and computed times. The closer the points are to the line, the better the agreement.

In summary, (x, y) graphs show the relationship between variables.

Bar chart

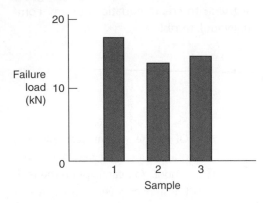

A bar chart simply compares values of one variable.

If you want a true visual comparison between values, the vertical axis should start at zero. You exaggerate the differences if the scale starts at a value greater than zero (as seen on many adverts).

Histogram

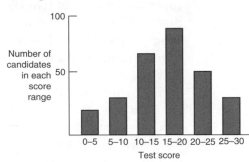

A histogram shows how data is distributed.

Pie chart

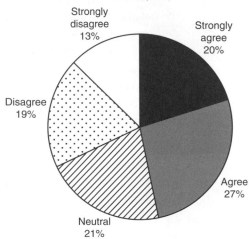

Responses to 'Existing arrangements are adequate'

Pie charts are good for showing how something is divided up (opinions for example). However, if the data simply indicates that 70% agreed and 30% disagreed, a pie chart is unnecessary.

Computer aids

Some of the data processing and presentation on your course will involve using software. This may be a dedicated package for specialist work or a spreadsheet that you have created or that has been provided for you.

Good knowledge of spreadsheets is very useful when analysing and presenting data. You need the skills to be in control. When you are working with a spreadsheet you must make decisions about how to present data for yourself; it shouldn't be left to default settings and wizards in your package. If you are not confident with spreadsheets it is definitely worth learning more. Listen to advice; a good source of guidance is the text book by Liengme (2015) – see References.

Realistic challenges

7

The bigger picture: Education and professional development

The content of an engineering or built environment degree is designed to provide a preparation for a professional career. The list of things you need to be good at while you are a student should agree fairly well with the list of things you need to be good at as a professional. They won't agree perfectly because a degree should be an education in its own right as well as a preparation for a profession.

This chapter is about the relationship between the requirements of the degree and the requirements of the profession. It seems best to do this separately for engineering and built environment.

7.1 Engineering

Your course

You should have access to a list of learning outcomes for each of your modules (or subjects). You should also have the learning outcomes for the whole programme (or degree course), contained in a document that may be called the 'programme specification'. A typical way of dividing up these outcomes is as follows:

1 Knowledge and understanding (of, for example, 'appropriate mathematical methods, including those needed for modelling engineering problems', or 'relevant principles of applied mechanics')
2 Cognitive (thinking) skills (to, for example, 'apply problem-solving skills in engineering applications')
3 Practical skills (in, for example, 'carrying out laboratory experiments and tests')
4 Transferable skills (in, for example, 'interpersonal and interdisciplinary working', or 'appropriate visual, oral and written communication')

For an engineering course to be 'accredited' (approved as the education component in a professional qualification) it must satisfy the requirements of the relevant professional body. In the UK the organisation that oversees this is the Engineering Council, and the requirements for course content are specified in a set of guidelines called 'UK Standard for Professional Engineering Competence', or UK-SPEC. The content of degree courses in any subject in the UK also has to comply with the appropriate subject benchmark statement published by the QAA (Quality Assurance Agency). Happily, for engineering, the defined learning outcomes in the benchmark statement are exactly those of UK-SPEC. So the same set of generalised learning outcomes are specified by both the academic quality agency and the central accrediting body. The task of accrediting specific degree courses is licensed by the Engineering Council to professional engineering institutions (like the Institution of Mechanical Engineers) but, again, this is all based on the learning outcomes in UK-SPEC.

UK-SPEC

UK-SPEC is partly about specifying learning outcomes for accredited degree courses, and partly about specifying what engineers who are finally seeking professional qualification should be capable of.

The learning outcomes for degree courses come under six headings (Engineering Council, 2014a):

- **Science and mathematics** (for example 'Knowledge and understanding of mathematical and statistical methods necessary to underpin their education in their engineering discipline and to enable them to apply a range of mathematical and statistical methods, tools and notations proficiently and critically in the analysis and solution of engineering problems')
- **Engineering analysis** (for example 'Understanding of engineering principles and the ability to apply them to undertake critical analysis of key engineering processes')
- **Design** (for example 'Apply advanced problem-solving skills, technical knowledge and understanding to establish rigorous and creative solutions that are fit for purpose for all aspects of the problem including production, operation, maintenance and disposal')
- **Economic, legal, social, ethical and environmental context** (for example 'Awareness of relevant legal requirements governing engineering activities, including personnel, health and safety, contracts, intellectual property rights, product safety and liability issues, and an awareness that these may differ internationally')
- **Engineering practice** (for example 'Awareness of quality issues and their application to continuous improvement')
- **Additional general skills** (for example 'Apply skills in problem-solving, communication, working with others, information retrieval and the effective use of general IT facilities')

When a course is accredited, apart from having discussions with staff and students, a team of academics and engineers will check that course content matches these requirements and that student work reaches the required standard.

UK-SPEC also defines what is expected of professionally qualified engineers, by setting a 'competence and commitment standard'. It contains useful summaries of the difference between incorporated engineers (who *'maintain and manage applications of current and developing technology'*) and chartered engineers (who are *'characterised by their ability to develop appropriate solutions to engineering problems, using new or existing technologies, through innovation, creativity and change'*). These word pictures may appear to be only subtly different but the key is that chartered engineers are associated with 'innovation' and 'new technologies'. It is generally accepted that chartered engineers must be strong in engineering analysis in order to solve completely new problems in an innovative way. UK-SPEC defines the 'competence and commitment standard' for the different levels of professional qualification. The main headings for incorporated and chartered engineers are given on Table 7.1 (Engineering Council, 2014b).

Table 7.1   *UK-SPEC competence and commitment standards*

Incorporated engineers must be competent throughout their working life, by virtue of their education, training and experience, to:	*Chartered engineers* must be competent throughout their working life, by virtue of their education, training and experience, to:
Use a combination of general and specialist **engineering knowledge and understanding** to apply existing and emerging technology	Use a combination of general and specialist **engineering knowledge and understanding** to optimise the application of existing and emerging technology
Apply appropriate theoretical and practical methods to design, develop, manufacture, construct, commission, operate, maintain, decommission, and recycle engineering processes, systems, services and products	**Apply appropriate theoretical and practical methods** to the analysis and solution of engineering problems
Provide technical and commercial **management**	Provide technical and commercial **leadership**
Demonstrate effective **interpersonal skills**	Demonstrate effective **interpersonal skills**

(continued)

Table 7.1 (Continued)

Demonstrate a personal **commitment to professional standards**, recognising obligations to society, the profession and the environment	Demonstrate a personal **commitment to professional standards**, recognising obligations to society, the profession and the environment

Professional qualification

When you have graduated, if you wish to become a professionally qualified engineer you will have a few years of supervised work experience to prepare you for professional review. The review is the assessment by a professional body to check that you have achieved the competence required. This is guided by UK-SPEC, in the form of the standards on Table 7.1. For example, an application to become a member of the Institution of Mechanical Engineers is supported by personal competence statements that demonstrate how you have achieved these standards in your professional experience so far (IMechE, 2014). The emphasis is on the professional and practical engineering aspects rather than the academic, because it is assumed that the academic requirements are in place as a result of your degree studies.

International recognition

Around the world the requirements for accreditation of engineering degrees are set by the relevant national bodies: the Engineering Council in the UK (as we have seen), Engineers Australia, Institution of Professional Engineers New Zealand, Engineering Council of South Africa, ABET (USA) and many others. Internationally there is some mutual recognition of standards for the accreditation of engineering education. For example the Washington Accord covers qualifications in a number of countries (including the UK, USA, Canada, Australia, South Africa and Japan) at the equivalent of CEng level, and the EUR-ACE System covers European accreditation of engineering programmes.

7.2   Built environment

Your course

You should have access to a list of learning outcomes for each of your modules (or subjects). You should also have the learning outcomes for the whole programme (or degree course), contained in a document that may be called the 'programme specification'. A typical way of dividing up these outcomes is as follows.

1 Knowledge and understanding (of, for example, 'building science with respect to materials, structure, services and internal environment', or 'how quality is achieved in construction processes and products')
2 Cognitive (thinking) skills (to, for example, 'analyse and solve construction problems of a technical and managerial nature')
3 Practical skills (to, for example, 'apply techniques used for the analysis and surveying of existing buildings')
4 Transferable skills (in, for example, 'interpersonal and interdisciplinary working', or 'appropriate visual, oral and written communication')

For a built environment course to be 'accredited' (approved as the education component in professional qualification) it must satisfy the requirements of the relevant professional body. There is quite a range of professional bodies representing the different professional disciplines within built environment, and we will consider some below. The content of degree courses in any subject in the UK also has to comply with the appropriate 'subject benchmark statement' published by the QAA (Quality Assurance Agency). Again, there are a number of these for built environment, for example:

Architecture
Architectural technology
Construction, property and surveying
Housing studies
Landscape architecture
Town and country planning

Subject benchmark statements define the nature and extent of the subject and set standards in relation to the knowledge, understanding, attributes and skills expected. If you want to find out more, have a look on the QAA website.

Professional bodies

Professional bodies have a big say in the content of degree courses. They also specify the standards that must be achieved after graduation for professional qualification. It's difficult to generalise about this across the built environment disciplines, so let's consider some examples.

The procedures of the Royal Institution of Chartered Surveyors (RICS) are called 'Assessment of Professional Competence' or APC. Three levels are defined. Level 1 ('knowledge and understanding') should come from degree studies, and the RICS would expect accredited degrees to cover the requirements. Level 2 ('application of knowledge and understanding') should come from the work environment, and level 3 ('reasoned advice and depth of technical knowledge') is likely to come from the experience of giving advice to a client.

All those seeking membership of the RICS should have the following 'mandatory competences':

Level 1	Accounting principles and procedures
	Business planning
	Conflict avoidance, management and dispute resolution procedures
	Data management
	Sustainability
	Team working
Level 2	Client care
	Communication and negotiation
	Health and safety
Level 3	Conduct rules, ethics and professional practice

In different professional disciplines there are different 'core competences'. For example, for building surveying at level 3 (RICS, 2015a), they are:

Building pathology
Construction technology and environmental services
Contract administration
Design and specification
Inspection
Legal/regulatory compliance

In contrast (for example) the core competences at level 3 for quantity surveying and construction (RICS, 2015b) are:

Commercial management of construction *or* Design economics and cost planning
Contract practice
Construction technology and environmental services
Procurement and tendering
Project financial control and reporting
Quantification and costing of construction works

As a further example, to become a chartered architectural technologist via CIAT, the Chartered Institute of Architectural Technologists, evidence of performance in seventeen competency areas is required (CIAT, 2010) to complete a Professional and Occupational Performance (POP) record:

Project inception	1. Client and user requirements
	2. Feasibility studies
	3. Sustainable development
Project planning	4. Project planning
	5. Health and safety
	6. Regulations

Design process	7. Concept design development
	8. Design proposals
	9. Technical design development
	10. Design information management
	11. Specifications
Construction management	12. Tenders and contracts
	13. Contract compliance
	14. Project completion
Professional practice	15. Management of meetings
	16. Professional relationships
	17. Continuing professional development

8
Personal development planning and recording

8.1 Personal development

You'll be at university 3 or 4 years, maybe 5 years, maybe more, and you'll be changing as a person all the time. There will be the big transition at the beginning which we've discussed in Chapter 1, there'll be the big transition at the end (more later); and in the middle there'll be transition all the time, probably both gradually and in bursts.

Does that mean more than one transition at a time? Well, that's up to you. When you start your course the most important thing is to negotiate the transition described in Chapter 1, getting used to university and university study. The big transition at the end is becoming a professional, but there is a long transition that leads to that point. After all you've got to actually get a job, and before that you've got to become the sort of person that an employer would want to employ. In Chapter 2 we looked at 'keeping an eye on where it's all leading' and that is the area of the middle transition, what some people call 'developing employability'. As I said in Chapter 2 there is no reason not to become totally involved in your student life, but you should just keep an eye on where your course is leading. It will probably make your course experience more interesting in any case.

I ran a scheme in which first-year students could be mentored by a practitioner from industry. It was a popular scheme with a good take-up, but it was not a mandatory part of the course and I was surprised that the take-up was not even higher. I asked a few students why they thought that was, and the general response was that many students in year 1 simply want to immerse themselves in student life, full stop. Plus there are coursework deadlines to worry about, and many other pressures.

I also organised the year-out placements at the end of year 2. Some students had been very active up to that point, taking advantage of the mentoring, maybe finding relevant summer work experience at the end of year 1. They were able to put together a very convincing CV and usually sailed through the selection process into a year-out placement. Others really worried about their CV (with good cause); there was no evidence at all of an active interest in the profession up to that point. They were very keen to take a year-out placement and really regretted not taking more opportunities earlier in the course. But there was a problem with their general awareness of the industry. They had no idea what sort of work they wanted or what they could expect. This was an obvious weakness in their CV.

Of course we're all different. Some of these students got their act together at the last moment and still impressed employers. But others didn't, and were not successful in finding a placement. And although they may easily have gone on to get a graduate job at the end of their degree, I found it impossible to avoid feeling they had missed an opportunity they would never have again, and I think they felt that themselves too.

So immerse yourself in being a student, and be more! Don't ever 'just be' anything. Have the energy and courage to take all the opportunities you can.

8.2   Reflecting, planning and recording

Reflecting on your development needs, planning how to satisfy them, and recording the resulting personal and professional development is a cycle. We look here at what it might entail while you are a student, and in the next section we look at what it means in professional life.

Why plan and record your personal development?

1   It is likely to be a part of your course and may easily be assessed and credit-bearing, as PDP (personal development planning).
2   It will help you get a good job.
3   It's a good practice that you will need to continue after you graduate.

When you are applying for jobs you will need things that make you different and stand out from other applicants. Recording personal development, with the consequence that you naturally develop awareness of what is expected, and raise your awareness of the industry, is a good way to approach this.

Reflection is an important aspect. Self-awareness is a very important attribute. In fact reflecting is a very natural thing to do. But often the process of writing it all down can seem artificial.

I'm quite reflective. I like to take long walks, thinking about things. But I know that lots of people aren't like this. Many engineers and built environment professionals think of themselves primarily as 'doers'. But that doesn't diminish the importance of reflection. Some of the most successful professional people are doers and very profound thinkers too.

Your personal development is likely to be recorded in a number of ways. It is likely that, as part of your course, you will be required to record an evaluation of your own skills and attributes in a format that is provided for you.

In an example from one course (Igarashi *et al.*, 2015), students are asked to consider their current performance against a detailed list of criteria under headings such as application of knowledge, team working, leadership and professional attitude. An example is given below. They rate these on a scale from 1 to 9 where 1 represents no expertise and 9 represents high expertise. This is particularly useful in indicating development, by comparing scores across the headings over time.

Application of knowledge	Score (1–9)
Can make rational arguments based on observation and proof	
Is able to organise knowledge systematically	
Recognises the absence of required information	
Can adapt standard practices to new and diverse situations	
Recognises opportunities to consider changes to accepted methodology	

As well as identifying your level of achievement you may need to identify evidence that supports the claim you are making. This will need to be genuine documentary or electronic evidence. This can be the hardest part for students even though it doesn't involve creating anything new. You need to think about the sort of evidence that would be appropriate and to make sure you keep any documents, files and comments that you think might be useful as evidence of your achievements in future. This applies to your time as a student, and to your career.

Another format you may be expected to use is one of your own creation. This will be a 'portfolio' (physical, on paper) or an 'e-portfolio'. You may be compiling this to satisfy a course requirement or it may be your personal collection of evidence to show potential employers. It is likely to include:

- overview statement
- your CV

- education transcripts and certificates
- evidence of learning on the course: design work, project work, etc.
- personal and professional development planning records from the course

- employment experience
- evidence of what you achieved at work
- possibly references or letters from employers
- reflective statement on work experience

- information about interests
- any relevant evidence (certificates of achievement, event programmes, etc.)

- more personal evidence of reflection and planning

Overall the purpose of all this reflection, planning and recording is to provide evidence of how you are developing, and to ensure that your development is providing you with the skills and attributes you need. This covers your course and everything else you do. For more advice see Cottrell (2010). It is good to align this as much as you can with the development needs you will have as a professional (as introduced in Chapter 7).

8.3   In professional life

We've considered the journey into professional life a little in Chapter 7. After you graduate, if you wish to become professionally qualified you will undertake initial professional development. You will be generating the evidence you need to show that you satisfy the criteria for professional qualification, and you may also have to show that you satisfy some institution-specific criteria as presented in Chapter 7.

Once you have gained your professional qualification you will continue with your professional development. CPD (continuing professional development) is expected of all professional people, to maintain and improve their professional competence. Apart from the obvious need for professionals to be up-to-date, there are legal aspects to this. If one of your decisions or designs was to be questioned legally it might be important to prove that you had maintained your CPD. Professional bodies monitor the CPD of their qualified members.

The records expected would show how you regularly reviewed your development needs, set objectives, planned how you would achieve the objectives, and recorded and evaluated the development itself. The development could be in absolutely any professional area, ranging from specialist expertise to strategic management, or from health and safety policy to presentation skills. It could take many forms, including a short course, attendance at a meeting, discussions with a mentor and learning from work.

As a student and a professional, you never stop developing, and there is always a need to record your development. PDP leads to CPD.

9
Practical work

9.1 The learning

In laboratories and workshops, you come into contact with the real thing: real equipment, real ways of working. You see physical phenomena first-hand, and can test important principles and theories with real data.

In my experience most students appreciate practical work, so I don't think I need to 'sell' you the idea. But I realise there can be a huge variety of experiences. It is sometimes very open-ended: you've got to work out what to do, learn by making mistakes, choose your own tests, work out what's going on. In others you are told exactly what to do and what to find out.

Access to facilities is part of this. Sometimes, for good reasons, it is only during the supervised class that you can use specialist equipment; in other cases you can go back any time to learn more for yourself; and other practical work isn't reliant on physical facilities at all.

Practical work is often done in groups, so all the issues of group work (to be considered in Chapter 11) come into play. I notice in the lab that some students seem content to 'just watch' even when the group size is quite small. It's such a shame. Things should be set up so that everyone needs to be involved, but regardless of that you should make sure you do get involved. You may never get a chance to do exactly this again – take full advantage.

Your records

Some practical work can be highly technical but many of the skills that you need while you are carrying it out are straightforward. One of the most important is to keep tidy, clear, well-structured records.

When students do an extended project with me in the lab, they usually keep excellent records. They have a lab book, they write in it neatly, they try to record everything relevant. Some students doing a short one-off lab class with me keep horrible records. Maybe it's not all their fault. We record a lot of data, and capture a lot of our observations now electronically. Also, when lecturers want students to capture a few key pieces of data from an experiment they often provide a printed format, or a ready-made spreadsheet. But the skill of recording data clearly on paper is still important, and if the practical investigation is one that you are creating for yourself, you will have to devise your own format for records.

So what do good written records look like? They are:

Tidy and well-structured – but not perfect. They form a working document.

Unambiguous – so that you or anyone else would understand each piece of data. So 'depth' is no good. Depth of what, when, how measured? A sketch may help to make things clear.

Comprehensive – nothing can be left out. Incomplete records are likely to be completely useless. This goes beyond data values that you know you must record. You should note possible sources of error, for example. Also, sometimes the most important thing is not quantitative at all, it is a change that you observe and must record in words.

Health and safety risk assessment

For any extended practical work that you are carrying out, a health and safety risk assessment is almost certain to be a requirement.

This will involve simple documents that record a logical thinking and planning process. In a laboratory, you should discuss this with the technician. Ultimately of course the aim is to keep you and everyone else safe. There will probably be a standard format for you to follow. It may be as simple as the example below.

Details of work	
Hazard	Precautions to be taken

Risk can be quantified and you may be required to do this. It is made up of two components: *likelihood* and *severity*. These are multiplied together to quantify *risk*.

likelihood × severity = risk

The likelihood of falling from a platform with guard rails is much lower than the likelihood of falling from a platform without guard rails. The severity of falling from a height of 0.3 m is much less than the severity of falling from 3 m. So the lowest risk would be from working on a platform with guard rails at a height of 0.3 m, and highest would be working on a platform without guard rails at 3 m.

In practice it is common to quantify risk using a simple numbering system. A typical example with a range of 1–5 is given below.

	Likelihood		Severity
1	Highly unlikely	1	Trivial
2	Unlikely	2	Minor
3	Likely	3	Moderate
4	Highly likely	4	Serious
5	Certain	5	Very serious

Clearly the lowest risk (likelihood × severity) would have a value of 1 and the highest a value of 25. The response to the estimated risk (for student practical work) might be as below.

Risk	Response
1–2	No special precautions required beyond normal safe working
3–5	Specific safety precautions must be implemented
6–10	Active safety management is essential
over 10	After consideration of safety precautions to reduce this risk, there may still be a case for not proceeding with the work as originally planned

In work that is inherently risky, like construction, 'residual risk' may also be calculated – the risk that remains after precautions have been taken.

A format for a risk assessment, including calculation of risk, might be as below:

Details of work			
Hazard	Possible consequences	Risk (= likelihood × severity) with reference to classification system	Precautions to be taken
(Example) Water on floor resulting from over-topping	*Slips on to the floor causing injury*	*2 × 2 = 4*	*Care to prevent over-topping Warning cones and mop readily available*

Writing up laboratory work

What you write as a result of carrying out practical work could be anything from a brief piece of analysis to an extended report. Advice on report-writing is given in Chapter 13, and on a major research project or dissertation in Chapter 14. There is such a thing as a 'traditional lab report', and you may be asked to write one (or several) of those. They are short reports following the principles of Chapter 13. There are standard subheadings along the lines of:

Lab report – main subheadings (alternatives in brackets)

Aims
Theory (Background)
Equipment (Apparatus; Experimental installation)
Procedure (Method)
Results
Analysis
Discussion
Conclusions

'Results' is likely to include data read directly in the experiment, and calculated values based on those. It is important to make it clear which are actually readings.

Here are some words to be careful about:

'Prove'   To really, seriously, **prove** that a relationship or law or principle is true would involve collecting a convincing set of evidence, and it is unlikely that your experiment has generated enough. However, you might be able to say that your results **confirm** something or are **consistent** with it.

'Compare'   You are often asked to do something like compare your experimental values with those suggested by theoretical analysis. To do this well, I suggest three stages:

1   Ensure that the two values are presented in a way that allows the reader to see easily what they are. This could be as simple as a small table containing just the two values:

Directly measured force	Force from momentum equation
3.1 N	3.3 N

However, for large data sets a comparison is normally clearer on a graph than a table.

2 Comment on the similarity or difference between the two values or data sets. This could be descriptive – slight difference, significant difference, etc. – but should be backed up quantitatively, for example by the value of percentage difference.

3 Discuss any difference (even if you are not specifically asked to). If there is a likely reason, explain it, for example 'Calculation of the force from the momentum equation included the velocity at the nozzle, but this will have reduced slightly before hitting the plate. This is an explanation for the directly measured force being lower'. If you have reason to believe that the agreement is affected by a specific source of error, explain this (see below).

'Error' Measurements are not exact. It is important to give consideration to measurement and observation errors when you consider your results. You can use your knowledge of the experimental procedure to estimate the likely error for different parameters, for example length measured by a rule, ± 0.5 mm. Error bars, which show this range visually, can be used on a graph (Figure 9.1), and these may even show how measurement errors have affected the extent to which data agrees with a theoretical relationship. Be specific about errors and explain them clearly. Any error that could be applied to almost any experiment should not be included. Don't use the word 'errors' as an excuse for a difference you can't explain. And never refer to 'human error'. This really just means that you messed up the experiment so the results are meaningless, and it would be better to say that, or preferably ask to repeat the experiment.

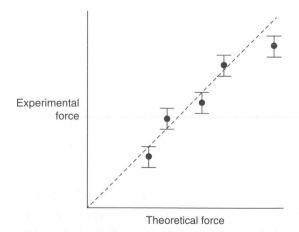

Figure 9.1 Graph with error bars. (The disagreement between experimental and theoretical values can be explained by the estimated measurement error for all but the largest value.)

10
Graphical communication and design

10.1   Visualisation and sketching

A significant element of the work of students and professionals in engineering and built environment is about turning ideas into reality. This chapter is about how ideas can be represented visually on their way to becoming real. It is typically a process in which ideas are visualised initially, then represented as sketches, then developed into detailed drawings or diagrams. The visual representation is the means by which the idea is recorded and communicated. The better the sketch or drawing, the better the communication. But what really matters is the quality of the ideas. And although sketches help in developing ideas, there is no sketch or drawing without ideas.

Visualisation

There is a significant debate about how much students should learn about computer-aided drawing packages on engineering and built environment courses. On several occasions I have been a member of a course team discussing this with representatives from industry on an advisory panel. Academics might assume that industrialists would want graduates (or placement students) to be highly competent with computer-aided drafting (CAD) packages so that they can use those skills immediately at work. But this was not the response we got from industry. Instead they told us that the most important skills were sketching and, in particular, visualisation: thinking clearly and effectively in 3-D. This was stated so strongly that we decided to include the word 'visualisation' in the title of the module.

Of course the evidence for good visualisation is in being able to produce effective sketches and drawings, and on being able to read drawings and to discuss and solve problems in three dimensions. So visualisation can't be separated from sketching and drawing. But it is the core skill that underpins the others.

Figure 10.1 Sections through an orange

How good are you at visualisation? What does a section through an orange look like (a) cutting through the segments, (b) in the plane of the segments? Do you agree with Figure 10.1?

Test yourself

Is a normal mirror a plane surface?
Is the top of a guitar plane? A violin?

When a sphere is intersected by a plane at any angle, is the intersected shape always a circle?

Name a three-dimensional shape that, when intersected by a plane, gives an ellipse.

Sketching

Everyone can sketch. People who say they *can't* sketch just mean they *don't* sketch or *won't* sketch. If you want to be able to sketch well, just do it. Keep a sketch book with you. Become the sort of person who sketches. Keep practising by doing.

There is much to be said for using a pen, not a pencil, to stop you worrying about 'being wrong'. Forget the idea that everything should be perfect. As you get better you'll find the sort of pen, and the sort of book, that you prefer. It's got to be right for you. But there's no need to be restrictive (or old-fashioned). A tablet instead of a sketchbook? Use what you're comfortable with.

Sketches often make better records of sights or experiences than photos. A sketch captures more than just the visual. In contrast, a photograph tends to capture less; many people, if they are taking a photo, simply don't *look*.

Sketching is a core skill. The better we are, the better we see our ideas, and explore our ideas, and the better we communicate them to others.

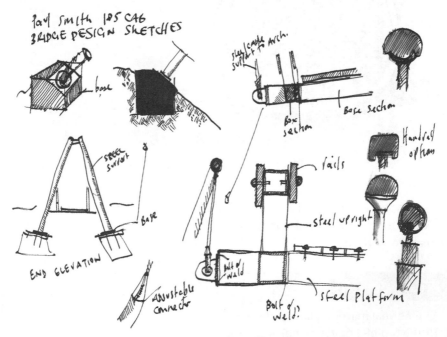

Figure 10.2 Developing design ideas through sketching

If you are sketching design ideas, the sketches and the ideas can develop together (Figure 10.2). Sketches can lead to ideas. Be general, open up. You don't have to include detail. Sometimes the detail in a sketch is in inverse proportion to the power of the ideas that the sketch represents. Great ideas can be represented by very simple sketches. Sometimes a sketch is an aide memoire, a step on a journey. Design is iterative. To produce detailed drawings, or design on a computer, you have to know what you are aiming for in advance. With sketches, you can explore ideas and let your aims evolve.

But OK, it's not just about spontaneity and ideas. There is a technique to sketching. The principles may be included in your course. Even if they are, and especially if they are not, you should learn more for yourself. There's plenty out there. You could start by checking out the work of Liz Steel and Francis D. K. Ching. There is also the Engineer's Toolkit within the amazing Expedition Workshed (expeditionworkshed.org).

Sketches are not just for the early stages in developing ideas. They can be used to communicate specific design ideas. In this case a sketch needs to communicate your intent clearly; you won't always be there to explain it yourself. And to take you further through the design process a sketch can be scanned into software and you can start modelling in 3-D.

Sketching and then working with a 3-D model really has benefits. While you are sketching you are limited only by your imagination, not by your knowledge of the software, but once you are working with the model you can start to appreciate the relationships between elements, spot potential clashes, and explore tricky corners that may be hard to visualise. Once you know what you need to investigate with the model there is then no reason why your knowledge of the software should hold you back – there is plenty of help in performing specific software tricks on the internet (or from your friends or via your course).

10.2   Moving to CAD

Drawings don't originate from CAD packages, they come from ideas that are visualised by someone. The thing that is visualised is in 3-D, but its representation is physically in 2-D, on a screen or a piece of paper. Software can help, but you must understand what is visualised.

To use CAD packages successfully in this context you must:

1   have reasonable mastery of how to use the software;
2   fully understand what you want to represent.

These are two separate requirements. You can't compensate for weak understanding of what you want to draw by having fantastic CAD skills. In any case a CAD package is not a design tool, it is merely a drafting tool, though it can certainly help you with details.

Here are some key things to think about when using a CAD package:

- Your drawings must be completely **unambiguous**.
- They must communicate **relevant detail**. For this you need to understand what detail is relevant and what is not.
- Think about **scale** and the appropriateness of your scale.
- Think about **line thickness**: it can be hugely effective.
- Establish clear **context**: a ground line for example.

BIM

In civil engineering, built environment and other branches of engineering that relate to building, building information modelling (BIM) is becoming a dominant concept. Let's be careful about what BIM actually is: it is not a type of software, it is a collaborative communication process. In many branches of engineering this approach has been the norm for some time.

As a process, the assembling of information about the design and functioning of a building is extremely complex. Imagine defining completely and unambiguously the process of making a simple cup of tea. There is more to it than you might think. Imagine explaining it to an alien, or someone who did not know what tea was, or water, or a cup, or milk or a lemon, or sugar, or heat. Scale this up to an engineering plant or a complex hospital building and try to imagine the number of elements that need to be defined and the complexity of their interrelationships. The team must design and develop ideas and concepts, and communicate and share these ideas across the team and beyond. Ideally this information should be held in a platform that can be assessed, poked and prodded, tested structurally and environmentally by anyone in the team at any time. That is the BIM model. It is fed information, and tweaked and tested by all interested parties; it is regularly updated and is accountable. It makes the information useful for the whole of the building's life.

10.3   Representing concepts and systems

We use diagrams for a range of purposes. We use them to represent how things are arranged (Figure 10.3) or how they operate (Figure 10.4). These diagrams of physical systems are not detailed drawings. They may look realistic but they do not give the precise details of what the thing really is. In fact detail is at a minimum, the minimum needed to communicate what is necessary.

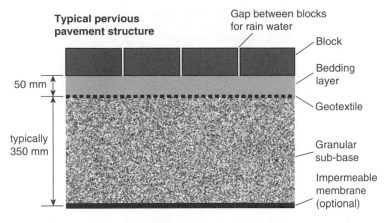

Figure 10.3 Physical system – how things are arranged

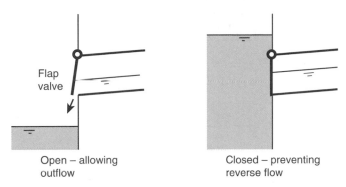

Figure 10.4 Physical system – how something operates

Diagrams are also commonly used to represent a physical system but in a conceptualised form (Figure 10.5). There is no representation of physical reality here, just of the concepts and the relationship between them. Diagrams can also show conceptual systems that are more abstract (Figure 10.6).

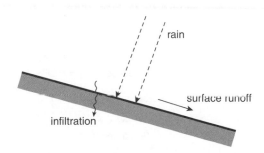

Figure 10.5 Conceptualisation of a physical system

Supporting the quality of the student learning experience

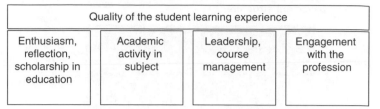

Quality of the student learning experience			
Enthusiasm, reflection, scholarship in education	Academic activity in subject	Leadership, course management	Engagement with the profession

Figure 10.6   Abstract conceptual system

Diagrams can show procedures (Figure 10.7) and control systems (Figure 10.8). Obviously if you study control as part of your course you will go into this in much more detail.

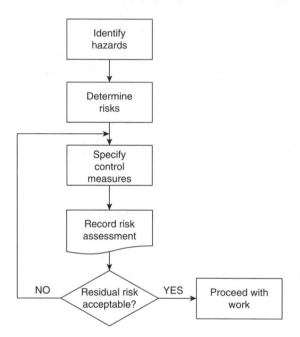

Figure 10.7   Procedure

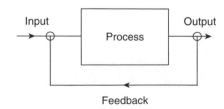

Figure 10.8   Control system

10.4   The essence of design

Design is one of the most important areas of skill in engineering and built environment. Many professionals in these fields describe themselves as 'designers' first and foremost. What this actually means in practice varies widely, for example in terms of the level of creativity, or the complexity of the technology being applied. Certainly design skills are highly discipline-specific and therefore not a suitable subject for in-depth treatment in this book. But there are some common aspects, and these relate to state of mind, to attitude and approach, and to professional development, and these aspects we will consider.

The essence is this. There is always uncertainty in design. There is never a straight line from the question to the answer. The uncertainty is an inevitable and fundamental characteristic of the design process, and it means there must always be an element of creativity in design. The good ideas will come from somewhere, but we don't know where that is going to be when we start.

I have always found it fascinating being with students with varied backgrounds as they approach a design task. Those who have some design experience, through work or previous study, settle into the task easily. They think about the brief, think about what they know and what they don't know, start to organise how they are going to go about the job. They are working and thinking hard, but they are comfortable with the situation. For them, the uncertainty is expected: it is routine.

Students without design experience are not comfortable. 'How are we going to do this?' they ask each other uncertainly. They feel the need to visualise a route that will take them to the solution, as they might with a complex mathematical problem, but because they don't know what that route is they are unable to make real progress, or even make a proper start.

So the essence is being comfortable with uncertainty, accepting it as the norm. You learn this from experience (of design on your course or in practice). And while you may be thinking 'in that case why do I need to read about this in a book?', the important thing, like so many aspects of personal and professional development, is to be aware in the first place of the need to develop the right state of mind for design, and to acknowledge and celebrate the progress you make.

I realise that if your course is fundamentally design-based, I am just scratching the surface. You'll pursue all this in much more depth on your course than I can here.

11

Team work and realistic projects

11.1   Team work at university

Working with other people is a fundamental aspect of most human endeavour, certainly professional practice in built environment and engineering. There are very few times when a practitioner is not working in a team. So it's entirely logical that university courses should include group work – as a vehicle for learning and developing team-working skills, and for the same reason there is team work in industry: that many real challenges are too big for one person.

But group work at university is a difficult area. We will look at this in some detail.

How do you feel about group work? My children (11 and 12) do plenty of group work at school and they both hate it. My son (at primary school) just gets frustrated with everyone else in his group messing about (though he probably ends up doing the same). My daughter (starting secondary school) quite likes leading a team, but only if everyone else does what she says. And even then she doesn't see why she can't take full responsibility and credit for her own ideas.

I'm mentioning this because I realise that you have probably had plenty of group-work experience before you started your university course. I expect at least some of the experiences were negative.

Some lecturers don't see the difficulty with group work. A pet hate of mine is when a lecturer fends off complaints about group work by saying, 'well, that's what it's like in industry'. This is only true in the sense that people work in teams in industry and all members of a team must demonstrate some team-working skills for a successful outcome. But there is a management structure in industry; few commercial organisations rely purely on the team-working skills of team members for efficient management. Plus employees at work are being paid, and are developing a professional reputation. You can expect your colleagues to behave professionally, and it is rare that they don't. In contrast, at university there is no commercially motivated management structure. Some teams may work very successfully. But some teams may be hampered by members who do not have the motivation or perhaps the skills to contribute fully, and this can severely hamper the work of the team. You hear about keen members doing the work for less keen members so the whole group will get a decent mark. In these cases some lessons may still be being learned about team working, even though these may be examples of projects that have been structured poorly by the staff organising them. But they present difficulties, and explain why so many students are cautious about group work. I asked some *part-time* undergraduate students (who work four days a week in industry) whether they thought the experience of team working at university was 'realistic'. Their responses are very interesting and are presented in Box 11.1.

Did they think that the team work experience, in terms of working relationships between team members, was realistic? There may be problems, but isn't that what it's like in the real world? There was some agreement with this.

> 'At work we don't get to choose who we work with. Some of the guys we work with are absolutely the worst people on earth – we'd never choose to work with them ... they're the cheapest, or they're the client, and that's who you have to work with, whether they pull their weight or not'.

They presented a mixed picture: not everything in industry is ideal in terms of team work:

> 'In industry you find that people generally do pull their weight – some don't'.

However, there is a management structure in the workplace: the boss can intervene if there is a problem in the team:

> '... when it gets really bad [at work] you have someone to step in'.

Interestingly, most conceded that this rarely happened. Many of the students said they simply sort out problems for themselves by talking directly to colleagues, even when they think the colleague is at fault: 'You'd just go and speak to him'. But then isn't that similar to the university experience? Definitely not:

> '... you have to act professionally at work'.

Team work in the workplace is moderated by the management structure, and also by expectations of professional behaviour and professional motivation:

> '...in industry ... you're all working on the same project, with a manager, getting paid'.

The attitudes of work colleagues are different from those of some of their team members at university:

> '[some other students] don't understand the consequences of falling behind.'
> 'If you're in the real world and you're working for clients, you can't afford to miss deadlines.'

Most felt that team working at university is not completely realistic:

> 'At university we only ever lose a mark or two ... if you make a mistake ... you've not cost your company a million pounds'.

Box 11.1   *Part-time students' views on team work at university (based on Davies (2013))*

Having said all this, group work at university can be very rewarding and be a vehicle for transformative learning and personal development. The purpose of this chapter is to help to make this true for you.

11.2   Good team work

Team roles

Most people prefer to take a particular role within a team. Typically, different people like to:

- take the lead
- come up with ideas themselves
- work with others to develop ideas collaboratively
- listen, and weigh up what they hear
- specialise in particular technical areas

Also, at particular stages in a project, specific skills of team members may come into play: writing, speaking, or picking up errors, for example.

> *It's important to determine strengths and skills of team members and allocate tasks that suit them. Not everyone does everything in a team situation.* 𝟃 **EMPLOYER**

The important point is that you need a balance: you don't want too many of one type and none of another. A team of leaders, or a team with no leaders, is not well balanced.

In some well-established systems of role definition you can find out your own favoured team role by completing a questionnaire. Or you may be required to identify your way of working on the basis of a role classification provided on your course for project work. Or, at the very least, you should think about the possible individual roles, and use these to help you and your colleagues form teams.

Forming a team

Team membership may be specified for you, or you may be free to choose your team mates, or there may be a structured process created by your tutors to support you in forming effective groups.

Where you have some influence over this I think the two main things to remember are:

1   this is important;
2   the key is *balance*.

Let's reflect for a second about what this might mean. You may want to work with friends. But remember that friendships can be broken or stressed in bad group work experiences, and the key remains *balance*. If it happens that your friends offer a genuine balance of team roles – that may work. But if you're all quite similar (a common characteristic for a group of friends) then working together as a team may not be successful. You can work with friends, and you can make friends through work, but if you don't have a balance of roles in a team, the team will not work as well as it could.

Working as a team

Here are some principles:

1 Leadership is not bossiness. The best leaders bring out the best in other people, they don't tell them what to do.
2 Don't confuse assertiveness (when you show and seek mutual respect) with aggressiveness.
3 Listen.
4 Be careful not to assume that someone who is quiet is not contributing. It is only when they're not listening that they are not contributing.
5 Keep asking, 'Is there balance? Is the team working as a team?'
6 Where there is a division of work make sure everyone understands what their share is, and record in writing your agreement about who does what.
7 When things go wrong, face up to what has happened and find a way of getting back on track.

How do you accomplish 5, 6 and 7? Answer – through 8:

8 Have effective meetings.

11.3 Effective meetings

People seem to love saying how much they hate meetings. I'm sure some of your lecturers say things to you like 'well, I'd like to make an arrangement to meet you then (at the time you have suggested) but unfortunately I have the delights of a department meeting to look forward to', or 'if I could add up all the time I waste sitting in meetings...' It's true that some official meetings are formal affairs that don't seem to get anywhere.

A professional team in the workplace needs meetings to share information and review goals, etc., and they usually have a very tight focus.

Meetings of a student team should be as focused as possible, and rather than use the language of formal meetings: 'chairing', 'agenda', 'minutes' etc., it would be better to concentrate on what is really needed to make a meeting of a student group effective.

Some group members may forget things, and some members may want to forget things, so having some material in writing is essential. A good starting point for a meeting is a record of the last meeting. There may be a lot to talk about, and jumping randomly from one topic to another will not be the most efficient use of everyone's time, so some sort of logical sequence for the discussion is useful. You can agree this at the start, though in that case you may waste time having a discussion about what you're going to be having a discussion about. One person could suggest a sequence for the discussion and it could be quickly agreed or adapted at the start. Then one person could make sure people stick to the plan during the meeting. A record of the meeting is needed as a starting point for the subsequent meeting, so someone needs to make that record. These roles for group members can be fixed or they can rotate. There may be requirements specified for your project in any case.

So there are some roles and some documents whose purpose is to make meetings efficient and effective. It's nothing to do with formal meetings, though there is some equivalence.

Meeting of student group	Formal meeting equivalent
Logical sequence for discussion	Agenda
Someone to make sure people stick to the sequence	Chair
Record of meeting	Minutes
Someone to make record	Minutes secretary

So what should be discussed?

Progress
The division of work and targets agreed for each group member should have been recorded in the notes of the last meeting. Near the start of the next meeting each member should report progress in turn. Have they achieved what was agreed?

Is the group on track?
This should be clear from the discussion of progress. If the group is not on track (is falling behind the progress needed to complete on time), some response is needed, for example adjusting some expectations, or shortening the period for the next phase.

Developing ideas

If you've got time, you can develop ideas during the meeting. But it would be best to make a decision about whether you want to have an open-ended discussion, or just to listen to each member's ideas to see if there is agreement.

New targets

This must be an outcome of the meeting, and the agreed targets will then determine the start of the next meeting.

11.4   Peer assessment

There are problems with group work, as we have discussed. In my experience built environment and engineering students understand why group projects are considered important, and realise they can learn a lot from them. The thing they really distrust is group assessment – in particular being assessed as a group, with one mark for all, when they are well aware that contributions by individual group members have not been equal.

'Group assessment' of students is not a meaningful phrase anyway. All assessment of students is individual. It is you as an individual that passes, gets an upper second, etc. The marks for group projects are processed in the end as individual marks. When staff talk about 'group assessment' they really mean individual assessment in which the assumption is made that it would be fair to give each individual in a group the same mark. And as we all know, sometimes it isn't.

Assessment is only group assessment when the final outcome is for the group. An example is team sports. One team wins, by so many goals, points, runs, whatever, even if some players have a much better game than others.

Staff could award individual marks in group work by making a judgement about the relative levels of contribution by individual members. They can also ask students to annotate the contents page of the report with 'who did what' – though that may not be fair because even if one individual wrote up the section on 'evaluation of alternatives' it is likely that they included ideas and input from other group members.

Maybe the differentiation between group members should be made by those who really do know who made the strongest contributions – the students themselves.

That is why you may easily be required to take part in 'peer assessment' in which you assess the contributions of everyone in the group. This may benefit you by increasing the fairness of the marking, and may also be something to learn from.

Feedback through peer assessment influences future behaviour through reflection on personal performance. ☌ **LECTURER**

If peer assessment is done well, it will give you the chance to assess your team members at an early stage in the project in a way that allows you and the others to respond to the assessment of colleagues. It should be done anonymously and may involve the use of software which helps with anonymity and the processing of the feedback and marks. Staff may also moderate the division of marks through scrutiny of materials like a reflective project log.

With peer assessment, staff assess product, students assess process. ☌ **LECTURER**

11.5   Learning from realistic projects

Realistic project work, usually involving team work, can be the most exciting, challenging, frustrating and rewarding part of your degree course. You may be lucky enough to encounter this type of work throughout your course, or it may be just towards the end – a sort of climax to your learning. The type of project will depend on your discipline; the potential list includes investigations, product development and testing, feasibility studies, design projects, development proposals, and many others.

All aspects will be challenging. You may be provided with lots of information, some may not be relevant (but this you will have to determine), some may change. You will need to think flexibly and not cling to assumptions. And, as we have discussed, you will have to find a way of working effectively as a team.

Some realistic projects require integration between different disciplines, with students working in multi-discipline groups. This can be challenging, but also mimics, and provides preparation for, professional activity. These are often described as 'scenario-based projects' – the whole project is based on a realistic, often real (direct from industry), scenario.

You are likely to be required to synthesise knowledge from other taught modules as well as finding solutions that involve integrating the different disciplines. Alongside the discipline-specific challenges are personal and professional development opportunities. The skills developed might be similar to those in Table 11.1.

Well-structured team-working projects may have quite a tightly specified structure. Each week teams may be required to meet to make specific decisions and plan for the completion of specific tasks, with staff available for consultation. Teams may be required to draw up a rota to outline who is responsible for

Table 11.1 *Typical skills developed in an integrated project (based on Austin et al., 2011)*

Technical skills	Personal and professional skills
• Clarify client requirements • Produce a detailed design solution which meets client and project needs • Undertake technical design as specified in the design brief • Formulate appropriate project management strategies for meeting specific client needs • Monitor costs from project inception through to tender • Produce relevant tender documentation	• Develop visual and verbal communication skills • Develop team-working skills in a multi-discipline environment • Demonstrate the ability to manage and control personal and professional development through reflective evaluation • Demonstrate good professional practice and employability skills

leading the group and producing minutes at each stage; each member may have to undertake each role during the project.

These projects are typically assessed through submission of technical tasks, through presentations, through submission of minutes etc., and quite possibly through peer assessment. Although tasks may have been completed by individual group members, the submission may have to be presented in the form of an amalgamated group report rather than a series of individual pieces of work. This often involves the group selecting and using a company name and logo and their work being formatted in a coherent professional manner, as would be expected in the workplace (Austin *et al.*, 2011).

These projects, especially in the final year of a course, can lead ultimately to exhibition at a public event including invited guests from the relevant professional practice areas.

These are exciting projects, and some students get carried away! It's important to do your best, but you may need to be careful that you do not spend so much time that it is bad for your other modules, or your health, or your family. Challenges are considerable, and overcoming them is a hugely valuable learning experience.

Intellectual development

12

The bigger picture: Learning and teaching

This chapter is an introduction to learning and teaching as seen by students and by staff. To some extent I suppose it's a behind-the-scenes look at things from a lecturer's point of view. I think it is useful for you to know what your lecturers are thinking. Also it's important and reassuring to realise that a lot of professionalism and enthusiasm goes into delivering your course.

12.1   Transfer v. construction

Some lecturers see teaching as simply transferring knowledge from themselves to their students. Actually I think all lecturers see it this way sometimes, myself included. I often find myself standing in the lecture room saying, 'Now, the key to understanding this is...'. I'm desperate to pass on my understanding, but there are at least two problems with what I am trying to do. The first is that we all learn and understand in different ways, and the key to *my* understanding may not be the same as the key to my students' understanding. The second is that, even If I back up my point with brilliant illustrations or by creating a topical industrial context, the whole approach is based on what *I* am doing. The students are just sitting. In fact of course the whole thing is for the students' benefit, not mine; it is to enable them to learn. So the focus should not be on what I am doing but on what they are doing. What the students need to do in order for them to learn is to construct their own understanding. If they learn the same way that I do and they are listening very carefully they may gain understanding as a direct result of my efforts, but there is no reason to assume that they will, and it is very unlikely that they *all* will.

Because of this, many lecturers see their role not as transferring knowledge, but as creating opportunities for students to construct their own understanding. This is the background to a number of approaches to delivering effective education.

Lectures

Traditionally lectures tend to be the fundamental building blocks in the time-table. They can be useful as starting points for learning (for you to construct your understanding). The lecturer, partly through personal enthusiasm, can 'sell' the importance, relevance and essential characteristics of the subject. There is quite a lot of evidence that high attendance at lectures (and other classes) is related to high performance in assessment, though opponents of traditional delivery point out that there is no evidence of *causation* (that high attendance leads to good performance); they say it is simply that the best students are conscientious and therefore attend, and also do well in assessment.

Enquiry-led learning

The term 'enquiry-led learning' covers a variety of approaches with different names, all of which aim to stimulate and structure student learning in a particular way. The best known is probably **problem-based learning**, PBL. (In fact some people use these initials to embrace *project*-based learning as well, though others insist that this is a distinct approach.) For some academics, PBL and similar approaches represent philosophical positions, strongly opposed to traditional lecture-type delivery, but in reality on most engineering and built environment courses both approaches exist, and can be used to complement each other. (Architecture is more distinct, as we will discuss shortly.)

The core characteristic is that the starting point for learning is a question, a problem, a challenge, not a set of pre-digested information imparted by the lecturer. The role of the tutor is facilitator, supporter of learning. You might say that the central principle is 'better a question without an answer, than an answer without a question'.

Most supporters of enquiry-led learning point to advantages in terms of the quality and depth of learning, but they also emphasise the development of professional skills, that graduates are ready for work. In vocational subjects this makes enquiry-led learning very relevant.

A potentially valuable component in these types of learning experience is input from industry. The most realistic problem is a real one, and often the most up-to-date and context-aware guidance comes from a practising professional. Not all practitioners are good teachers, but many are excellent and really enhance the learning experience.

Education in architecture and some other design disciplines tends to be studio-based. Students work on design projects and specific challenges in a studio environment with support from tutors, and learn through discovery and through having their ideas challenged. It certainly has a lot in common with the enquiry-led learning model but it is so well established in architectural education that it is generally seen as a distinct approach.

12.2   Assessment

There is a widely held belief that there is too much assessment on many university courses. 'So why don't they do something about it then?' I hear you ask.

One reason is that there is a desire to give students, especially early in the course, early feedback on their performance so they can correct misunderstandings and have a clear idea of expectations. Another is the commonly held belief among lecturers that 'students won't do it if they don't get a mark for it'.

But there is a general move towards more emphasis on 'formative assessment' – which has the aim of simply providing constructive feedback, and less emphasis on 'summative' (final, mark-contributing) assessment. Examples of formative assessment are solving problems and asking questions in an 'examples class' or 'tutorial', completing online quizzes, or a review of design or project work in advance of submission.

12.3   Lecturing staff

Who are the lecturers on your course? They should be people with a strong understanding of their subject. This can be gained through engagement with the subject as a researcher, or through application of the subject as a professional practitioner, or preferably both.

Lecturers with a strong research background have a lot to offer. That is partly because through their research experience they are indisputable subject experts, and this gives their teaching credibility: they know the limits of their subject; they know how the subject is being developed. It is also because as researchers they are learners themselves, and know what it is like to have their knowledge challenged – not in exams, but in the competitive world of publishing and attracting funding. Their research is likely to be motivated by enthusiasm and excitement for the subject and this should come across to students.

There is of course great value in experience of the real world. It is nice to be taught by lecturers with industry experience, who know what things are really like. For these lecturers, it is enthusiasm and excitement for real-world challenges that will come across.

Of course, one thing that lecturers with a research background and a practice background have in common is that much of their work is actually education.

Indeed it is common for new lecturers to be required to take a course in learning and teaching – effectively a teaching qualification. The attitude towards these courses among lecturers in vocational disciplines like engineering and built environment is a bit mixed. Young research-based academics often resent having to spend time away from research, though in fairness many new lecturers are inspired by the ideas they come into contact with and generate some excellent ones of their own. One of the problems is that good practice in education is heavily related to the nature of the discipline, and generic educationalists may see things differently from subject specialists.

13

Reports

In university study and in professional practice you are likely to write plenty of reports, and it's certainly true that the practice you get as a student is a good preparation for work as a practitioner. Some reports are short and some are long. The content of this chapter is relevant to all reports, though length and purpose do affect content and format. An important report for students – a major research project report or dissertation – is covered specifically in Chapter 14.

13.1   Defining the purpose and readership

You can't write something if you don't know what it's for or who it's for.

Here are some types of report you may write while you are a student:

Test/investigation report
Research report
Project report
Proposal report
Feasibility study report
Design report
Specification
Manual
Survey report
Costings report
Visit report
Fieldwork report
Training report
Work experience report
Professional development report
Progress report

For student reports there are two types of reader. One is the reader that your brief has asked you to address, the client perhaps, and the other is the lecturer who will be assessing your work. You've got to pretend to be addressing the first, but you need to make very sure that you have the second in mind when you write.

Here are some types of readership you may be asked to write for:

Technical
Non-technical
Funder
Client
Community
Contractor
Approving authority
Employer

13.2 Structure

It is the structure of a report that is its strongest characteristic. In contrast, when you read a novel it's quite different: the structure may not be obvious until you have finished the last page. You may have to read several chapters before you even know for certain what it's about. An essay is different again: it should have a clear structure, but the structure is embedded in the writing.

In a report the structure is very obvious; it should be the first thing you see, communicated by the title, by a clear structure of subheadings, and by content that tells you straightaway what the report is about.

A report of any length should have components similar to those in the sample in Box 13.1.

The numbering system is not only logical, it makes the structure explicit. It helps the reader to see straightaway what the report covers and how the content is set out.

Flood management at Pindleford, Devon – an appraisal of existing arrangements

Summary
Contents
List of figures
List of symbols

1   Introduction
      1.1 Background
      1.2 Aims
      1.3 Objectives

Box 13.1   *Sample report outline*

2   Site details
    2.1 Location and catchment
    2.2 Community and industry
    2.3 Development plans
3   Flooding data
    3.1 Flooding history
        3.1.1 Floods of 1922
        3.1.2 Floods of 1954
        3.1.3 Flooding events since 2000
    3.2 Current predictions
        3.2.1 Extent of 100-year flood
        3.2.2 Critical impacts
4   Current arrangements for flood management
    4.1 River wall
    4.2 Berm
    4.3 Temporary storage in park
    4.4 Flood wardens and emergency arrangements
5   Appraisal
    5.1 Criteria and grading system used
    5.2 Grading for each component
6   Conclusions and recommendations
    6.1 General conclusions
    6.2 Recommendations

References

Appendix A Historical records
Appendix B Photographs of existing features

Box 13.1   (*Continued*)

13.3   Components of a report

Summary

A short report (less than about 10 pages) may not require a summary. Longer reports should have a summary right at the start, on the first page that follows the title page. The summary (also called Abstract or Synopsis) should summarise the entire contents. It's not a scene-setter (covering only the first part of the report), it's a brief representation of the content of the whole report including the aims, and the conclusions and recommendations. For most reports the summary should be between half a page and one page in length. The summary may be seen separately from the report itself, though you also have to account for the fact that some readers may read the whole report and not the summary. So the summary, and the report itself, must both say everything, but the summary must say it in far fewer words. The summary can be written last, in fact that's probably the best time to write it, but you should avoid cutting and pasting from the main report to produce the summary, because it can be annoying for a reader who has taken the trouble to read both the summary and the whole report to read exactly the same sentences twice.

A sample summary is in Box 13.2.

An 'Executive Summary' is slightly different. This is longer than a normal summary (two pages maybe) and is for someone who will not be reading the whole report.

Summary

Pindleford in Devon is susceptible to flooding because of its location immediately downstream of Dartmoor in a steep valley with old properties close to the river and a historical bridge. There were serious floods in 1922 and 1954, and there have been several less severe floods in recent years. A variety of measures for flood management have been introduced and this report presents an appraisal of these. In the context of the community, industry, and plans for development, the report presents the flooding history of the area, and identifies the critical locations that would be affected by the current prediction of the 100-year event. The main flood management arrangements currently in place are the river wall, an artificial berm for increased river flows, the facility for water to flood further into parkland, and emergency planning including supervision by volunteer flood wardens. Criteria for the appraisal, including a grading system, are set out, and the current arrangements are graded. General conclusions are that the arrangements are fit for purpose and the main weakness identified is that in the case of a flood that exceeded the capacity of the temporary storage in the park, although it is clear that flood water would overflow to an existing road, there is not enough protection for properties lying on the route. The report recommends providing additional protection for these properties.

Box 13.2   *Sample summary*

Contents page

The contents page would look like Box 13.1 except it would include the page numbers for each component (so the pages of your report must be numbered!). Obviously the contents page directs the reader to where the content is located, but in a report it does more. It says, 'Look – this is what is in this report, and this is how it is structured'.

Introduction

Often a report needs some brief statement of the context or the problem before the aims and objectives can make sense.

Aim

This is where the reader finds the answer to the important questions 'what is this report for?' and 'why should I read it?' An 'aim' is an overall direction; it is often expressed in just one sentence.

Objectives

If the aim is where you want to go, the objectives are how you get there. Objectives are milestones on the way to achieving the aim.

A typical introduction to a report would be like Box 13.3.

---

1 Introduction

1.1 Background

The River Pindle carries runoff from Dartmoor through the Devon town of Pindleford. The steepness of the valley has meant that historically development in the town has been close to the river, and the main bridge, built in 1765, constrains the river flow in flood conditions. The town has experienced some serious flooding. In 1922 three people were killed, and in 1954 there was significant damage to property. Flood management measures have been introduced over time, and have been tested several times in recent years. In 2009 some further works were carried out, and emergency plans put in place.

1.2 Aim

This report presents a study of flood management at Pindleford. The aim is to carry out an appraisal of the existing arrangements and where appropriate recommend improvements.

1.3 Objectives

The objectives are:
- to identify the main characteristics of the location, the upstream catchment, the community, its industry and plans for development
- to present data on flooding history
- to ascertain critical impacts of the currently predicted 100-year flood
- to review current arrangements for flood management
- to appraise these arrangements using pre-determined criteria and a grading system
- to present the resulting appraisal
- to present general conclusions and recommendations for improvement

---

Box 13.3 *Sample introduction*

Main content

The main sections of the report present what has been done and how it has been done. As we have discussed, the structure of the content should be immediately clear to the reader.

The style should be as we discussed at some length in 4.4. If you're about to write an important report, it would be worth looking back over that section now.

If your report is presenting the results of a research project there is more specific guidance on content in Chapter 14 (14.7). In any report with academic content you will need to follow the conventions of referencing and citing, as covered in Chapter 5 (5.4).

In many engineering and built environment reports, the communication of information is a joint effort between words, tables and diagrams. The words guide the reader through the report, but where appropriate they direct the reader's attention to a table ('as given in Table 5.2', for example), or a diagram, usually called a 'figure' ('represented in Figure 6.1'). When readers turn a page they look straightaway for the next piece of text to read, even if there are also tables or diagrams on the page. So in a report all tables and diagrams are specifically referred to in the text. You should make sure you do this consistently in your report. It's easy to forget because in other types of writing, newspapers and magazines for example, this is never done. You don't read about a famous victory in sport that 'fans celebrated well into the night, as illustrated in photo 3'.

In a report of any length the numbering system for tables and figures is based on the section or chapter numbers. In section 6, the tables are Table 6.1, Table 6.2 etc.; and the diagrams are Figure 6.1, Figure 6.2. But you do not use the sub-section numbers, so the tables in sub-section 6.1 are not Table 6.1.1 etc., the tables are simply numbered sequentially right through section 6. In a report of a few pages, it is more common simply to number the tables as Table 1, Table 2 etc. When you refer to tables and figures using a number system in this way there is no need also to write 'as shown *below*', or 'on Table 2.6 *above*'.

Sometimes tables and figures come thick and fast. That's fine. For multiple tables you can just write 'this data is shown in Tables 3.3 to 3.9'. Where you have masses of data or loads of graphs, it may be neater to place them in an appendix. If the reader would have to turn two or three pages full of diagrams halfway through reading a paragraph, an appendix might be the answer. The disadvantage is that the reader then has to hunt around in the back of the report to find what you are talking about. But a way round that can be to give some sample data within the chapter and the rest in an appendix.

Tables and figures should have captions. Captions just provide a bit of additional help to the reader. They also, as introduced in 5.4, provide a tidy way of citing a source for a diagram, for example 'Figure 3.9 – Schematic representation of choices available (based on Thompson (2015))'. Don't be put off if sometimes a caption feels a bit pointless. If your text says 'the relationship between Q and h is given on Figure 4.5', the logical caption for Figure 4.5 is 'Relationship between Q and h', and that is still helpful. Though in fact it could be even more helpful if you reminded the reader of the definitions by making the caption 'Relationship between flow-rate (Q) and water level (h)'.

Conclusions and recommendations

These are the outcomes. In industry this is the bit that matters most and so this section is often presented at the start of the report. In education it is more common for it to be at the end. This is where the reader sees how the aims and objectives have been achieved.

The conclusions usually start with a bit of context but this is brief.

Sample conclusions and recommendations are given in Box 13.4.

---

6 Conclusions and recommendations

6.1 General conclusions

This report has described flood management at Pindleford, Devon, and has presented an appraisal of the existing arrangements based on a grading system reflecting level of protection (historical and predicted), impact on the environment and community life, sustainability and cost of operation and maintenance.

The existing river wall has been evaluated as fit for purpose and generally well designed and maintained. Some lengths of the wall consist of temporary barriers only; this approach has been used to protect views of the river, and across the river, during times of no flood risk. The temporary barriers are easy to erect and no problems have been experienced. The berm has been assessed as coming into operation for the one-year flood and this is consistent with observations by the community. For more severe floods where the capacity of the berm is exceeded, the parkland beside the river is deliberately flooded, enclosed by the wall around the park. This arrangement has been in place since 2009 and has been known to operate twice in that period with no problems of damage or public safety. However, if the capacity of this temporary storage is exceeded it is clear that exceedance flows will follow a course down an existing road (Bank Road). On this road existing properties are at risk.

The system of voluntary flood wardens is exemplary, and has helped to harness community support for the flood management arrangements generally.

Box 13.4 *Sample conclusions and recommendations*

6.2 Recommendations

The main recommendation is that additional protection should be provided for properties on Bank Road that lie on the route that would be taken by exceedance flows from the temporary storage in the park. These should include raising kerb levels, assessing properties for temporary protection, and providing additional protection to properties in Cedar Close by improvements to the existing wall. No major changes to the other physical arrangements for flood management are recommended. The positive role of the volunteer flood wardens should be recognised wherever appropriate, and support given if necessary to succession planning.

Box 13.4 (*Continued*)

13.4   Presentation quality

Quality of presentation is important. But that's definitely not the same as saying 'it doesn't matter what you say as long as it looks nice'.

So what does this really mean? That your report must be fancy or eye-catching? No. For a report in engineering and built environment, the key is that the presentation should be *professional*. This will help to build confidence in the reader's mind. You can find out about professional presentation by looking at any kind of report in the public domain: produced by trade or research organisations, professional institutions, or by your university, for example. Also, students with professional experience have a good idea of what's needed to produce a professional report from their experiences at work, so get some ideas from them if you can.

Remember that I'm referring specifically to presentation here. The actual content and structure of your report may not be at all similar to a professional report. Reports by commercial organisations often present answers to questions or solutions to problems, and the readers are really just interested in the answer or solution. In a university report, even in a vocational/professional subject, the focus is not just the answer but how it was derived. Commercial reports sometimes seem to say 'here's the answer (and if you want to know how we found it look in the appendix)', whereas a university report is more likely to be structured to say 'this was the problem, this is how we analysed it, and this is the solution we came up with'.

A lot of presentation is simply skill with font, layout and graphics. No one is expecting you to have the skills of a full-time document designer, just to use the skills you do have to achieve a professional standard. Remember that the structure of the report must always be explicitly clear. Make sure your presentation emphasises the structure and never obscures it.

Make sure you use all the information on presentation provided on your course. Follow the conventions of your profession, and your department (standard cover sheet, for example).

Appropriate binding depends on length. A few pages don't need binding along the spine, but more than about 10 sheets do. Make sure the binding allows everything to be seen and accessed easily, including sheets that are not A4. Don't encase paper unnecessarily in plastic. Unless you are told otherwise, print double-sided because it saves paper, looks good, and is often expected.

13.5 Summary of the characteristics of a good report

Before you submit your report you should have a reasonable idea of whether it's good or not, but here are some specific pointers.

	In a good report, this is:	For guidance, see section:
Structure	explicit and helpful	13.2
Writing	free from distracting misuse of words and punctuation	4.1
Style	simple and clear while professional and formal	4.4
Introduction	where the reader finds the answer to the questions 'what is this report for?' and 'why should I read it?'	13.3
Conclusion	where the reader sees how the aims and objectives have been achieved	13.3
Presentation	professional	13.4

14

Research methods

14.1　Student as researcher

You are likely to carry out a significant research project, or 'dissertation', towards the end of your course. This chapter is about the research methods you may need to use, and about writing up your research.

What is research for?

Cheap flights
Research is finding something out. So finding the cheapest flights for your holiday is research. Some people actually use the words, 'I'd better research the cheapest flights before we book'.

Solving real problems
Background research for solving a problem in professional practice might mean finding information about the properties of a material or an unusual product on the market. Similarly, background research for a university assignment might involve reading around a subject, being sure of your facts. This knowledge is new to you, and might be used by you in a particular way, perhaps even a unique way, to solve your problem or complete your assignment, but the knowledge itself is not new. It existed in the book or on the website (and in the minds of the people who wrote or presented it) before you read it.

Academic knowledge
Serious academic research involves making a contribution to knowledge, working at the limits of understanding of a specialist topic, identifying gaps and extending knowledge. This is genuinely new knowledge. Undergraduate students are not expected to push back the boundaries of knowledge but they are expected to know something about those boundaries. Students at level 6 (Bachelor's

degree with honours) are expected to have 'coherent and detailed knowledge, at least some of which is at, or informed by, the forefront of defined aspects of a discipline' (QAA, 2008). At level 7 (Masters) 'much' of their knowledge needs to be 'at, or informed by, the forefront of their academic discipline'.

Undergraduate investigatory project

An undergraduate investigatory project has to be more than just finding out something that is new to you but is not new in any other sense. And yet it does not have to involve pushing back the frontiers of knowledge. So what should it be? The word used most commonly is '**original**'. Anyone looking for the cheapest flight to Prague on 8 August would be asking the same question and would get roughly the same answer. Anyone researching porous asphalt would have access to the same information; but if they used the information in the novel solution of a drainage problem, although the knowledge would not be original, the application might be. Anyone working in the laboratory on a relatively specialist topic would almost certainly be acquiring a unique set of data and would therefore be able to conduct original analysis or comparison with theory. If existing software is being used to study unusual case study data, it is likely to produce original results.

   An important distinction in this context is between primary and secondary data:

Primary data	Data that is collected by the researcher
Secondary data	Data that already exists

How to do well in your student research project

Before we get more deeply into the nature of a research project, here are some general suggestions for the approach you should take to your own project.

Plan

Whether or not you are required to produce a plan for your project anyway, it is definitely worth planning your time carefully. Remember to include (or blank out) periods when you may not be able to work on your project.

Use your supervisor

Your supervisor is there to guide you. There may come a point when you are seeking guidance and your supervisor may say, 'well, I'd like you to think about that for yourself'. It's your project, not theirs. But when you feel you need guidance – ask for it. All the best project students I have supervised have come to see me a lot. I've never thought, and certainly never said, 'oh, not you again'. The project students I see less are the ones that haven't been doing much work and don't want to admit it. No one wants that.

Manage your time

This is to do with planning, of course. But during your project be careful to keep to your plan, and update it when necessary. Be clear about priorities: you must balance your project with your other work. If you feel you're getting stuck with your project, don't put it to one side, get yourself unstuck (maybe see your supervisor).

Be clear about your aim

As we will see, everything relates to your aim. It may change as your project progresses, but always be clear about what it is. Memorise it; be ready to tell anyone what it is.

Work efficiently but accept there will be dead ends

At the start of the project you're bound to do some dithering, but then the pace picks up and you start delivering. Find a way of working that gives results. Be confident; get involved; get the job done. Occasionally you will waste time on an idea that goes nowhere, but don't be frustrated about that. You had to find it out. You may even be able to include it in your write-up to show what you did and to save other people from going down the same dead end.

Look after your data

Store your data carefully. Organise your sheets, your files, your electronic files. Give your data the respect it deserves; it's your vital resource.

Record your thoughts

Some people insist you should have a research log or diary. In some types of research it is considered an essential part of the process. In any case you should record your thoughts and care for them like your data. You'll think so hard about your topic during the project that you're bound to come up with some good ideas. But you may be so busy or tired that you won't realise you've come up with something valuable. Also, your ideas may develop as your project progresses and it is important to capture this development. So record your thoughts all through your project and keep them safe.

Think for yourself

Have the confidence to value your own ideas. It's your project. Your supervisor guides. *You* make the breakthroughs.

Plan your report early and write as soon as you can

Start with a list of chapter headings, then sub-headings. It will all have a format like Box 13.1. Do this early on. When you can write something, do it. Show it to your supervisor. The sooner you get feedback on your writing the better. The best time to write about your procedure is while you are still following it.

Don't aim to write your literature review in one go. As soon as you've read some interesting sources, write about them. You can always tweak it later. Keep a log or table of references from the start, with a summary of the main points. Don't worry about whether you will use them all in the end. It's better to keep them all than lose the ones you need.

Leave plenty of time for writing up
Plan backwards. How long to print, to bind? Before that, how long to get comments from your supervisor? Before that, how long to finish the final bits of writing? If your final report is poor (usual reason: too much hurrying and not enough time for checking) your mark won't be particularly good, however brilliant your work and ideas. What a waste!

Enjoy the challenge
Research projects are always hard work and throw up some big challenges. But they're *your* challenges and you really prove what you can do when you overcome them. Overall the experience can be very satisfying and the report or dissertation you produce is tangible evidence of your achievements, ready to show employers and others.

14.2   Components of a research study

Most research studies, whether based on experiments, modelling, surveys or whatever, include the following components. I am deliberately considering them without a specific context so you can start to think about what they might mean in your own study.

Research aim

You need to define your aim early on, though there's nothing to stop you adjusting it later. It should be brief, one sentence if possible. But it should also be specific. Try to make the wording of your aim as distinctive as possible. After all, if your study has originality there should be something unique about your aim.

Aim: not specific	The aim is to investigate xx.
Aim: specific	The aim is to investigate the effect of xx on xx in conditions where xx is xx.

Together with your aim, you will have objectives. Your aim defines where you want to go, and your objectives define how you will get there. There is more on this, with examples, in 13.3.

An alternative approach to clarifying the purpose of your research is to specify your **research questions**. This is more common in social science and education than engineering and built environment, but it is a useful device. It forces you to think specifically in terms of the questions you are trying to answer.

Research question	What is the effect of xx on xx in conditions where xx is xx?

It is good to have one main question and a few sub-questions.

A more specifically scientific approach is to start a research study with a **hypothesis** (plural: hypotheses). This is a proposition that is tested by means of the study. The hypothesis (in the same way as an 'aim') determines all the other aspects of the study: methodology, experimental design, etc. The outcome of your study will be either to confirm or disprove your hypothesis. This defines a study in a very precise and scientific way but is not so suitable where there are 'grey areas'.

Awareness of the literature

OK, your project has originality, but what has been done before that is relevant? What must you know so that you don't reproduce someone's work or repeat their mistakes? Here you could be considering other research projects, other student projects even. It is important that you identify any gaps in knowledge to demonstrate that your research will add value. Awareness of the literature will deepen your knowledge of critical subject areas that relate to the aims and objectives.

Methodology

Methodology is the type of approach used: experimental, case study etc., and how the approach is applied in your specific study to achieve your aims. There is a wide range of approaches, and we will consider them shortly. You may already know which type of approach you are going to follow. The key here is that whatever your methodology may be, you must define it, and then follow it consistently. If your research is to determine what proportion of men wear hats, your methodology could be observation of a sample. You could identify your sample as anyone you saw as you walked to the campus by your normal route over five days. Your methodology could be criticised, indeed you would be expected to evaluate it yourself, but you could still present your findings as the outcome of clearly defined research. What you couldn't do is take a diversion past the police station on day 3 because you were worried that you hadn't seen any men with hats.

Ethics / Health and safety

Where research involves human subjects (even if it's just asking them non-personal questions), research needs ethical approval. Your university will have procedures for this. If there are no human subjects you may still need to actively confirm this. The other side of this, which becomes more significant for experimental work, is health and safety approval.

For a questionnaire or an interview on a low-to-medium-risk project (where there are no personal issues), the key is 'informed consent'. Your respondents must know what your project is about and how their responses will be treated, and they must record their consent. For interviews, participants must be made aware that the interview will be recorded. This usually involves a 'participant information sheet' explaining what the research is for, how they will be involved, and who to contact if there's a problem. It's normal to explain that the responses will be used in analysis and may be quoted. For interviews, respondents need to sign a form that gives their consent. For questionnaires the information sheet can explain that it is assumed they are giving their consent by completing the questionnaire.

We have covered health and safety risk assessments for practical work in Chapter 9.

Data

Your data is everything you have observed. Numbers, words; velocities, voltages, responses to questions; notes you have made. (By the way, some people insist that 'data' is plural, so I should have written 'Your data **are** everything you have observed'. No one is right or wrong, though it's probably best to do what people expect just to keep them quiet.)

Analysis

This is a bit like defining methodology. You have a complicated set of data: how do you decide what it means? You must define your approach to analysis. Sometimes the approach may seem self-evident. You will look at your measurements, plot them against something, and see if the relationship agrees with theory. But even here you are defining your approach. With more wordy data this is harder. You may need to create a scoring or coding system in which you give values to particular responses, for example. In this case you must define the system, use it consistently and evaluate it. There is no case in which you just look at the data and magically declare what it means.

Outcomes

Conclusions must relate to your aim, or provide answers to your research questions.

Research keywords

Here are some keywords. The context for most of them will become clearer in the following sections but I thought it would be helpful to give the explanations in one place.

empirical	based on observation and measurement, not on theoretical analysis
inductive	building up a theory from evidence
deductive	starting with a theory and confirming it through observation
grounded	findings derived from the data without preconceived ideas
quantitative	using numerical data
qualitative	using descriptive data
subjective	based on a person's perceptions
objective	independent of any person's perceptions
accuracy	the closeness of a physical measurement to the true value
precision	the repeatability, stability and constancy of particular measurements
sensitivity	the extent to which small changes are detected by measurement
validity	that results represent what you claim they represent
reliability	that results are repeatable
respondent	a human subject taking part in a survey
reflexivity	awareness by the researcher of all the reasons why they might be biased
triangulation	confirming a finding by seeking it more than one way

14.3   Studying physical phenomena

14.3.1   An experimental study

An experimental study could be in a lab or in the field. We have covered some basic elements of practical investigations in Chapter 9.

In an experimental study we are usually trying to find something out about the physical world by isolating it, or miniaturising it, or reproducing it. You set out what you are trying to find out in your aims. The next stage is to consider how you are going to find it out.

Experimental design

Design of your experiment may not be completely up to you. In a lot of student research projects there are constraints. You may have to use the gear that is already in the lab, and you may need to follow protocols that have been established in previous or continuing work. There is a huge range of possible types of experimental study; important aspects of experimental design, whether or not you are in a position to determine them, may include the following:

Equipment required; design of installation
Accuracy needed in measurements
How data will be captured
Range of cases to be studied
How the parameters should be varied
The cases that will be studied
Preparation of specimens
Materials required
Limitations of the lab experiment in representing the real thing

The purpose of the research is to achieve the aim, so the experimental design must have a close relationship with the aim. Assuming that the aim is precise (as it should be) then the relationship between the aim and the experimental design must be precise too. Of course the aim can change, but you should make sure there's always a secure and precise relationship between the aim and the experimental design.

When students want me to comment on an experiment they are proposing for their project, I always ask first, 'what is your precise aim?'

Design of your experiments is likely to be in conjunction with your supervisor and the staff in the lab. In a study that is being carried out in stages, the experimental design in later stages may relate to the results of earlier stages.

Let's say you're going to carry out a research project on using waste materials as a cement replacement in concrete. You are drawn to the project because you know that cement production is associated with high carbon dioxide emissions, so it would be good to use less cement in concrete and replace it by a material that would otherwise be discarded. That affects the properties of the concrete, so how much cement can we afford to replace? Your reading and your discussions with your supervisor suggest that the realistic range for this particular replacement material is 15 to 40 per cent. Your supervisor has advised you to study seven different concrete mixes with varying levels of cement replacement. One must be a control mix, with no cement replacement. You decide to look at three other mixes at first: with 15, 30 and 40 per cent replacement, and otherwise identical. (The 'otherwise identical' part is important: you cannot draw meaningful conclusions from comparing two cases that have more than one difference.) Once the results from these tests are known, you can concentrate the remainder of your tests on the part of the range that is most promising. What will you test for? There are several properties that might be important in different applications, but you decide that compressive strength is the most relevant. (It's clear from your reading that this would be top of the list, and your supervisor advises you to concentrate on just one property.) You will make the mixes with guidance from the laboratory staff, and, in compliance with the relevant British Standard, you will test standard cubes after 7 and 28 days. All the equipment needed is already in the lab.

That, in brief, is the basis of a simple experimental design. It all needs to be planned carefully. You spot that the last mix must be made 28 days before the last test, and there must still be time to analyse the results, finalise writing the report, and so on, before the submission date.

The number of times a particular test needs to be repeated depends on a balance between the statistical methods you may be using in your analysis, and the time and resources needed for each test. This is something you must discuss initially with your supervisor and the staff in the lab.

Data collection

So now the work starts and you are conducting your experiments, taking measurements and capturing data, in accordance with your experimental design.

One of the most important aspects is the setting up, checking and, where necessary, calibrating of instruments and any data logging system. You must know exactly what your data represents in terms of physical properties. You must check for zero errors. Often you can't put right any problems after the event. Never just hope for the best. Check, confirm, recalibrate when you need to.

Apart from quantitative data, your own descriptive observations may be essential. Record these in words or by sketches or photos. All this could be vital to help you make sense of your results, and photos can help with your write-up.

Data analysis

When we analyse data, we are trying to make sense of it. We are looking for underlying patterns, for significant relationships, or maybe checking for agreement with an accepted theory. We are doing this in order to achieve our aim.

A lot can be learned from simple plots. We may be making sense of the data by doing this, or we may just be 'getting to know' the data a bit more. When you start acquiring data it is good to plot it in some simple way as soon as you can. Don't expect some great 'light bulb moment' straightaway, just think of it as exploration. It's a mistake to gather all your data and only then start the analysis. Starting the process early might (for example) mean you can identify a change you need to make to your procedure before you've wasted too much time.

You can plot by hand on graph paper, or from a spreadsheet. If your data has been downloaded from a data logger on to a spreadsheet it obviously makes sense to use the spreadsheet to plot it, and if you have good spreadsheet skills you can do a lot of exploration this way. But you have to do the thinking. There is no spreadsheet wizard for answering the question 'what is my data telling me?'

Here is an example of a bit of exploration by plotting. There is no such thing as a typical example, but I hope you can see the general approach I am trying to describe.

You have carried out some experiments that involve measuring the velocity of particles moving in water flow in an open channel. You have taken measurements for two particles. One is slightly less dense than water and it floats on the water surface; the other is slightly more dense than water and moves close to the bed of the channel. You have studied conditions with water flowing at a series of different depths.

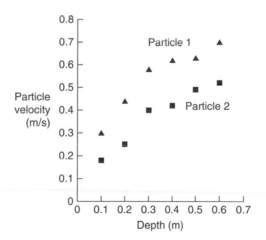

Figure 14.1   Data as points

The data is plotted as points on Figure 14.1. It doesn't look very smooth but that might be because of some errors that you are aware of. If you simply join up the dots on a spreadsheet you get Figure 14.2. Does that show anything valuable? There's a 'kink' on the line for particle 2 (the denser particle) at a depth of about

0.4 m, and for particle 1 (the less dense) at a depth of 0.5 m. Is there an explanation for this? Nothing obvious.

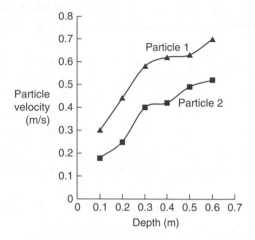

Figure 14.2 Simply joining the points

What about the mean velocity of the water? You would expect the floating particle to travel slightly faster and the sinking particle slightly slower. The water mean velocity has been added to Figure 14.3. That helps. Lines with this shape do seem to fit the particle velocity data (Figure 14.4). You would expect all the lines to pass through the origin. That affects the shape of the lines, though these lines have only been drawn for the measured data.

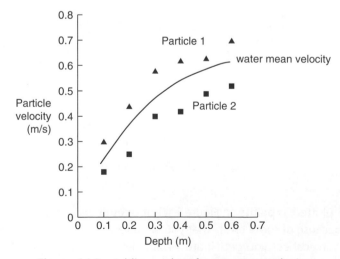

Figure 14.3 Adding a plot of water mean velocity

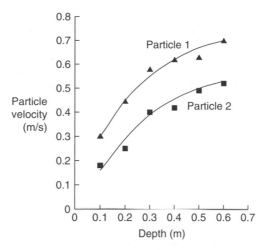

Figure 14.4   Lines that might fit the data

All this is suggesting that it might be interesting to plot the particle velocity as a proportion of mean velocity against depth (Figure 14.5).

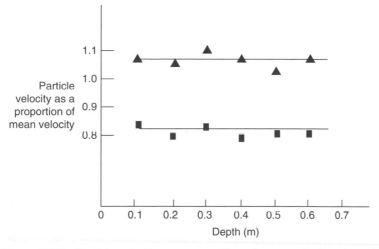

Figure 14.5   Particle velocity as a proportion of mean velocity plotted against depth

Interesting! Lines at fixed values of particle velocity as a proportion of mean velocity fit quite well, so perhaps there is a constant value for particles with different densities? But two particles are not enough to confirm this. Back to the lab!

To take possible errors into account when analysing data we can plot data with error bars as shown in Figure 9.1.

Sometimes a point is so far from the general relationship that it is appropriate to disregard it as an 'outlier'. Of course we must be very careful not to disregard data just because it doesn't fit with what we are expecting. But at the same time there is no point in being distracted by a data point that is probably erroneous. Ideally we should take more measurements or repeat the experiment to be sure one way or the other.

Working in the lab

Experimental research can be very satisfying. Working in the lab is a great opportunity to learn about a new environment. Make the most of it: embrace the practical side of what you're doing; enjoy being in the lab.

If you need to have something made, or to have some equipment adapted, or if your experiment needs special instrumentation or specialist support, lab technicians may become the most important people in your project. I've worked with some great lab technicians and seen students work very well with them. Occasionally students are unsure at first, but they quickly learn to work with technicians as partners. Your commitment is important. Make sensible arrangements and always stick to them. If you can't help changing an arrangement about working in the lab, give plenty of notice. Communicate technical details unambiguously.

14.3.2   Computer modelling

Aims of research projects involving computer modelling

The overall purpose of a computer model is to represent, or 'simulate', reality. If we have confidence in a computer model we can answer 'what if' questions (what would happen if a system is exposed to particular conditions?) without having to carry out physical tests or field measurements. In some cases computer modelling and physical testing are alternatives: for example to find out the effect of particular conditions on a complex engineering system. This ability to simulate special cases is behind a lot of modern advances in the way things are designed. Prototyping can be based entirely on computer simulations even in complex structures like space vehicles.

Alternatively, some research involving computer modelling is to develop the software itself. This could be for a model that is already in development, in which case the aim would be to expose the software to new conditions. Or it could be for a completely new model, in which case the aim would be to establish appropriate representations of reality within the model and test their effectiveness.

Software	Typical research aim	Outcomes
Well established	To test 'what if' cases in preference to physical testing	Better understanding of the phenomena being studied
In development	To further develop the model by exposure to new cases	Better understanding of the modelling approach used
New	To develop a completely new model and test its effectiveness	The ability to model new cases

Confidence in a computer model

The application of established finite element software to structural systems is based on a firm understanding of the mechanics of the system. The modelling approach is well established and understood, and the package may be supported by years of development and upgrading. The modeller still makes decisions that affect the accuracy of the simulations (for example the fineness of the mesh), but provided the input data is reliable and comprehensive, there can be a good level of confidence in the output. Another area of modelling where software is well established is computational fluid dynamics (CFD).

Models that predict some environmental systems can be subject to random variations in natural conditions, and the physical processes are so complex that some degree of simplification in the model is inevitable. Here a high level of confidence is less appropriate. It may be necessary to use data measured in the actual system to **calibrate** the model (adjust the value of certain model parameters to ensure good agreement). This means that some physical measurements may be required for effective simulation. The resulting simulation may then only apply to the specific conditions for which the calibration is valid.

Comparing physical test results with simulations

Some student research projects involve comparing the results of physical testing with the results of computer simulation. This allows originality (there's a good chance that no one will have carried out an identical comparison before) and provides opportunities for in-depth exploration of both physical experimentation and computer modelling.

One thing that needs some care is how you define your aim. If the software you are using is new then your aim is likely to be related to the development of the model. You are seeing how successfully the software performs in representing particular physical cases. But be careful that you do not think of your experimental results as 'correct'. You must still be careful about identifying experimental errors, or reasons why your experimental results do not completely represent the real thing. If the software is established modelling software you must be clear about what you might be discovering. If there is good agreement between the simulations and experiments, are both correct? Or are both equally incorrect? If there is disagreement, is there a way of knowing which is more wrong?

Examples

Let's look at some specific examples of research carried out by students. We will see how computer modelling is being used to find something out.

Crack failure of a pavement overlay (established software)

This work used established finite element modelling software to study the behaviour of a special concrete layer added to the top of a damaged road surface. The software was used to simulate the effect of a rolling load on the surface, which is also a physical test used in the laboratory. Figure 14.6 shows the output from the model. The different shades indicate stresses within the road layers. A crack has formed in the layers of the pavement and the model demonstrates that part of the added concrete layer would become detached. A separate study had involved comparing the computer simulations with laboratory results and this gave sufficient confidence in the computer model that it could be used to predict the structural behaviour of the layers in the pavement.

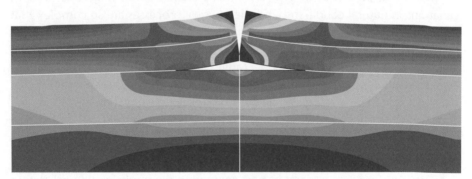

Figure 14.6　Output from finite element model

This study involved using finite element modelling software to answer 'what if' questions about a physical system.

Dispersion of solids in sewer systems (software in development)

This work used a computer model that was in development. It had been tested for a number of scenarios in which the results were easy to predict by theory, so there was some confidence that the main components of the model were working effectively. The model showed how solids, carried at different proportions of the water velocity (heavy solids more slowly, light solids more quickly), would be distributed in storm flows in a sewer. One set of results is given in Figure 14.7.

In Case A all the flow enters at the upstream end of the pipe. The graph shows how the concentration of solids would vary within a storm wave when it reached a point 2000 m downstream. The solid line shows the distribution if it is assumed they all travel at the water velocity. The dotted line shows how they would be distributed if it is assumed they travel at different velocities. There's a clear difference. However, for Case B, where inflows are distributed evenly over the 2000 m length, there is very little difference.

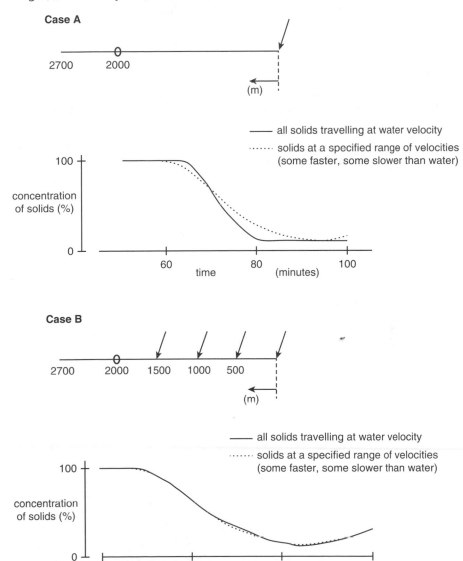

Figure 14.7 Comparing simulations

This study involved no comparison between the results of the computer model and experimental or field measurements, but does suggest something worth knowing: that for sewer systems with a number of distributed inflows (the normal case) there may not be any need to represent the different velocities in this way in a model of solid dispersion. By comparing like-with-like cases we're finding out something about how models represent reality.

Note the use of the words 'suggest' and 'may'. With just one set of results we haven't really 'shown' this to be true, and certainly not 'proved' it.

Erosion and deposition of solids in sewers (new software)
This work involved a new model, and the aim of the study was to determine how well the model represented reality. Artificial solids were placed in an actual length of pipe and then sudden inflows were introduced to see if the solids were picked up and, if so, how long they took to travel a length of 1250 m. Figure 14.8 shows one of these flow waves (solid line representing flow-rate at the upstream end; dashed line at the point 1250 m downstream). The small circles represent the actual solids: the time they were inserted (open circle) and, having been picked up and carried by the wave, the time they reached the end of the study length (filled circle). The behaviour of the solid as simulated by the computer model is represented by the squares. For a simulated solid inserted at the same time as the actual solid (open square at the same time as the open circle), the simulated time to reach the downstream end agreed quite well with the measured time (filled square at a similar time to the filled circle).

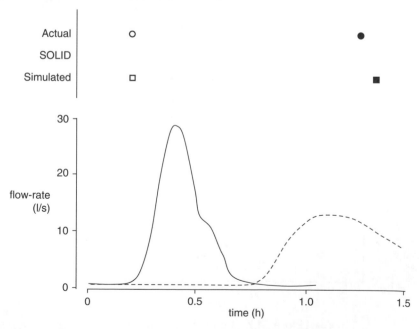

Figure 14.8 Comparing simulated and measured results

So this study involved comparing the results of model simulations with actual field measurements, and provided some confidence that the model was successful in representing the physical processes.

14.4   Studying human factors

The approaches we will consider for studying human factors are very different from those we have considered for studying physical phenomena.

Let's consider an example. There is a water-saving device (called 'Flowpal') that progressively constricts the flow of water to a shower or bath once a certain volume of water has been exceeded. It can be used in 'training mode' in which the specified volume is gradually decreased over a period of weeks.

There's no point in a device like this if people won't use it, so a research project will investigate public acceptance. Would people use this device? If not, why not? Answering the first question will involve collecting and analysing numerical data (quantitative research). The second will involve descriptive data (qualitative research).

14.4.1   Quantitative research

The research questions here might be:

- what proportion of people express a willingness to use this device?
- do people willing to use the device have particular characteristics?

The most common tool in quantitative surveys of human factors is the questionnaire.

Questionnaire design

An extract from a questionnaire to investigate public acceptance of Flowpal is given in Figure 14.9.

Several different formats are used here:

1   Straightforward tick box as in question 1.
2   'Delete as appropriate', as M/F in question 2.
3   A space for entering data, like a number (age in question 2), or text (the water-saving devices in question 3). The text input is an example of how questionnaires can collect qualitative as well as quantitative data. It is more time-consuming to process text data than simple ticks.
4   Response across a scale, as in question 4. This 5-point scale (called a 'Likert scale') is a common format.

This questionnaire is about saving water in the house, especially in showers and baths.

1. Does your home have a water meter?   **Yes** ☐   **No** ☐   **Don't know** ☐

2. Please tell us about any children who live in your home   Child 1 **Age .... M/F**

    2 **Age .... M/F**   3 **Age .... M/F**   4 **Age .... M/F**

3. Do you currently use any water-saving devices? **Yes** ☐ **No** ☐ **Don't know** ☐

If Yes, please name them ....

....

Flowpal is a water-saving device that progressively constricts the flow of water to a shower or bath once a certain volume of water has been exceeded. It can be used in 'training mode' in which the specified volume is gradually decreased over a period of weeks.

4. Please indicate your level of agreement with the following statements

	Strongly disagree	**Disagree**	**Neutral**	**Agree**	**Strongly agree**
If installed in someone's home, this device would save water	☐	☐	☐	☐	☐
I would like to save water in my home	☐	☐	☐	☐	☐
I would like to have this device in my home	☐	☐	☐	☐	☐

Thank you for completing this questionnaire.

Figure 14.9 Extract from questionnaire to investigate public acceptance of Flowpal

It's very hard to get questionnaires right. There is nearly always some ambiguity that you didn't spot but someone else will. It is definitely worth 'pilot testing' your questionnaire: trying it out on some people first as a check before using the questionnaire on the actual sample. You should record how you developed your questionnaire as a result of the pilot test.

Sample/procedure

How will we find a sample of people to complete our questionnaire?

If we were studying how engineering companies responded to something, we might identify companies in our area, try to make contact with them, and then perhaps visit or send the questionnaire in the post or electronically.

An alternative is to have the questionnaire available on a website and provide potential respondents with the link. This is a good option for many types of survey. A number of external organisations provide this service; the responses are held for you until you are ready to analyse them. This is usually all free of charge, though you usually have to pay for any serious analysis. But check first – your university may not allow you to use some of these services.

The problem with our survey is that no one will have heard of Flowpal and if we want to know the general public response to the concept we will need to actively recruit respondents. To ensure a reasonable number of responses we would need face-to-face contact, and capture their responses by asking the questions directly. The whole idea of Flowpal will need proper explanation, so people fully understand what they're being asked about. Going up to people in, say, a shopping centre, and asking if they would answer a few questions about saving water, might take them too much by surprise. Where would the idea of a water-saving device be more in context? Maybe outside a big DIY store. How might we attract their attention initially? Maybe by having a big display about water-saving. But wait, that might only attract people who are interested in saving water anyway. We need responses from a more genuine cross-section of people. We are interested in responses from people with children, so maybe balloons are the key! Kids can't resist them. 'I want a balloon Mummy, come and talk to this person'.

Quantitative analysis

In the end 135 questionnaires were completed. The responses to question 4 (last part: 'I would like to have this device in my home') are given on Table 14.1.

Table 14.1 Summary of responses to 'I would like to have this device in my home'

	N (total 135)	Strongly disagree	Disagree	Neutral	Agree	Strongly agree
With meter	84	31 (37%)	25 (30%)	18 (21%)	10 (12%)	0
Without meter	51	47 (92%)	2 (4%)	2 (4%)	0	0

The actual number of responses is given, though clearly percentages are needed to compare responses from those with and without water meters. That comparison is shown on Figure 14.10.

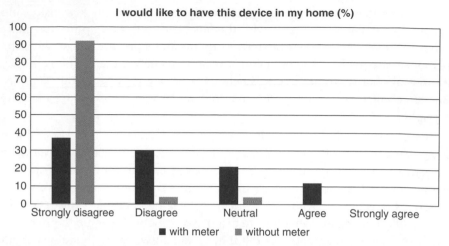

Figure 14.10   Plot of responses to 'I would like to have this device in my home'

It seems that the general response to the idea of using Flowpal is not positive, and it is significantly less positive for those without meters (who have no financial incentive to save water) compared with those with meters. The word 'significantly' is used in a descriptive sense, but 'significance' has a precise meaning in quantitative analysis. Suppose another question had asked how many showers the respondent took in a typical week, and the results had given a mean of 6.1 for those without meters compared with a mean of 5.4 for those with meters. Is that 'statistically significant', or is it just a chance result? A statistical test can be carried out to determine the probability that this result is just chance. A good source of information on these aspects is Rowntree (2000) – see References.

14.4.2   Qualitative research

The outcomes of the quantitative research above indicate that positive responses to use of this device are disappointingly low. But why is that? There are some indications from the quantitative study but we need to go deeper.

The research questions now might be:

- how do people react to the idea of using Flowpal?
- what is it about this device that is / is not attractive to people?

So qualitative research aims to find out how people think and feel about things. This is done most effectively through interaction. In an interview the researcher interacts with people by asking questions and giving prompts. In a focus group the researcher gets a group of people together, and by providing prompts, encourages them to interact with each other.

Interviews

Structure

Interviews are classed as structured, semi-structured or unstructured. In a structured interview there is a set of questions to be asked and the researcher is not permitted to deviate from the wording or the order of the questions. This might suit a survey being carried out by several interviewers where it was important that the questions were always the same, or where some special 'surprise' was embedded in the interview and it was important that it was revealed in the same way to every respondent. In a semi-structured interview there is a basic design: a list of questions or topics, but no constraint about order or wording. It is helpful to back up each topic by a list of prompts to make sure that no opportunities for getting the complete picture from each respondent are missed. Unstructured interviews are unscripted interactions between the researcher and respondent. The researcher has specific aims but tries not to place constraints on where the interaction goes. This approach requires special skills on the part of the interviewer and is unlikely to be appropriate for a student project in built environment or engineering.

Interview design

Let's assume that we are planning a semi-structured interview relating to the water-saving device. Here is a list of questions and prompts that could be used to guide the interview.

Topics / questions	Prompts
1 Concerned about water usage?	
2 Water meter?	Know how much you pay roughly? How do you feel about it?
3 Children?	Genders Ages
4 Do you try to save water?	How? Use any particular devices? Why / why not?
5 Describe Flowpal, what it does, what impact it would have; rough idea of monetary savings. How do you feel about it?	Try to get a full picture in relation to each point made by the participant.
6 Would you use it?	Why / why not? Try to get a full picture in relation to each point made.

Conducting the interview

Interviews must start with some scene-setting. The participant must be given the information sheet and asked to sign a form to confirm consent. You must make an audio recording of the interview, so the recording device must be switched on. The first question or topic should be a warm-up to start the participant thinking about the general area of interest. It's probably better if the interview design does not have precise wordings so that what is actually said by the interviewer can be as informal and inviting as possible. The interviewer should listen attentively to encourage the participant to respond fully, and prompt where necessary. If the participant says anything relevant or interesting the interviewer must pursue the point to get a full response. It may be enough just to say 'perhaps you could say a bit more about that'. You don't want to play back the recording later and think (too late) 'why didn't I ask more about that?' You should keep your questions and prompts brief; only talk yourself as much as needed to get the participant talking. Don't express your own opinions and ask whether the participant agrees.

Interviewing comes more naturally to some people than others, and if you are unsure about your skills it might be worth doing 'practice interviews' on people you know first.

Individual or group

Individual interviews can achieve depth in probing individual responses and may be easier to arrange. Group interviews can be valuable and may be an efficient way of involving more respondents. There can be scope for some discussion in the group, and for seeing if there is consensus among the respondents. However, the structure would be similar to an individual interview, with the interviewer taking the lead in guiding the discussion and probing responses.

Focus groups

Focus groups are quite different from interviews. There may still be a structure, but the focus is not on the interaction between the researcher and the participants, it is on the interaction between the participants themselves. Numbers between six and twelve are normally recommended. The researcher would have some questions or comments, and often also some physical materials, images or a display for example, to stimulate discussion, and would then only provide a steer or prompt where needed. Focus groups basically tell the researcher what is revealed when a group of people get together to discuss something.

Analysis of qualitative data

The first stage is the hardest work – transcribing the text from the recording. If you are new to this type of research you really should create a full transcript. This should give every word spoken by your respondents, and accuracy is important – there may be something about a particular phrase used that becomes important during the analysis. But it does take a long time – I usually estimate that it will take ten times as long as the actual interview to create the transcript. For focus groups it can take even longer because more than one person may be speaking at a time. If you are using software to support the analysis you can work direct from the audio, but that means learning how to use the software. As you become more experienced you may be able to transcribe the parts of the recording that are most important and summarise the rest, though you can only do that if you are certain about what is most important, and you may simply not know.

Then you must analyse the transcripts from all your interviews or focus groups. There are many ways of doing this, but I will suggest here a simple approach which I will term 'thematic analysis'.

Your data is the responses that you have recorded. You need to make sense of this. One fundamental principle is that you should not assume that the structure you should use in your analysis is the structure you used in your interview. Some people may have answered some questions very directly but others may have said something similar, or something contradictory, in response to a completely separate question. So you must look for structure in the responses as a whole, not as separate answers to questions.

There are three stages:

1   Coding

This involves using a coding system to identify topics within the responses. This can be done in an iterative way, rough codes first, then refined to identify common threads. An example of rough coding is on Figure 14.11. It gives some extracts from the transcripts for four respondents, with the initial coding in the right-hand column. All these respondents had water meters.

1 I would like to save money on my water bill, actually I would like to save water full stop, sort of saving the planet if you like, and I'm sure that the amount of water we use in the shower and the bath needs to come down a bit, but I don't like the idea of finding the water coming through the shower slowly getting less. Even if it was gradual and you could still finish the shower as you say, that would be annoying. I'd be – give me a break, I'm just having a shower – I'd just want to switch the thing off and stop it making life difficult.	*Wants to save money* *Wants to save water per se* *Doesn't like idea of losing control* *Annoying* *Would want to switch it off*

Figure 14.11   Example of coding

2 I think I could limit the length of my showers. Maybe if there was an alarm or something that went off when I'd used too much water. I don't like the idea of suddenly the water goes down. OK it keeps flowing for a while, but it's out of your control. But for the kids, yeah, I can see it making a difference. They'd just give up if the water flow went down a bit, it wouldn't take much, they'd lose interest. And for the bath, yeah, they really waste water in the bath.	*Simple warning as alternative* *Doesn't like idea of losing control* *Would work with children* *Children: bath especially*
3 No, really, no, really, I pay for my water and I decide. I work hard and I want to enjoy a shower when I get home. It's worth it. If I want a long shower, that's like buying anything else: you buy it because you want it. It's typical actually, I'm sick of it. Don't do this, don't do that, you're doing this too much, you shouldn't do that. We know better.	*Hostile to idea* *My money, my choice* *Resentment, people who think they know better*
4 I was first thinking 'why not have something that warns you when you've used too much water', 'it's time to stop' sort of thing. But I do want my water bills to go down and up till now I haven't done anything about it. I suppose this is sort of handing over the responsibility. This won't let you do anything but save water in the end. My kids – the strange thing is they're always on about the environment, well when they're bringing back stuff from school anyway, but I think they mean it. But they are the ones that waste water. One of them (he's at primary school) had to make a water diary – how much water he used every day. We worked out how much water there was in the bath and it came to what a person should use in a day – a whole day – just one bath. I can't persuade him to use less water. 'My sister has the bath up to here', he goes. Well, this will sort him out!	*Simple warning as alternative* *Welcomes loss of control* *Children waste water (even though they are environment-aware from school)* *Children: bath especially* *Hasn't been able to change children's behaviour before*

Figure 14.11   (*Continued*)

2   Identifying themes

The coding can be used to find common threads. I can't really illustrate this fully with what we have here because only extracts have been given. But it looks as if one theme might be children wasting water, and another might be loss of control.

3   Discussing the data using the themes

This type of data is usually presented in a descriptive style. It would be inappropriate to put emphasis on numbers, by continually using phrases like '7 out of the 13 respondents indicated that they would be prepared to...'. An overview of the responses should be given, and it is common to illustrate this by presenting quotes from the interviews. It is important not to overdo the quotes. The reader of your analysis is expecting to read your analysis, not a scrapbook of quotes.

> Several respondents said they felt their children were the most likely people in their home to waste water, and that this system might be helpful in limiting that waste.
>
> *'But for the kids, yeah, I can see it making a difference. They'd just give up if the water flow went down a bit, it wouldn't take much, they'd lose interest'.*
>
> It was felt that children were particularly wasteful with water when having baths.
>
> *'And for the bath, yeah, they really waste water in the bath'.*

14.4.3   Quantitative and qualitative approaches together

I have described this study of Flowpal as starting with a quantitative study. To find out why the quantitative data suggested that take-up of Flowpal would be low, I have suggested that a qualitative study would be carried out to go deeper into the issues. That is common, though the opposite order is sometimes used. In this case the interviews might have led you to place particular importance on children and that could be developed in the questionnaire design. Using both methods together allows **triangulation** by comparing and overlaying the findings of the different methods to add validity to the results.

14.5   Case studies

Many student research projects are case studies. It can be helpful to identify your project as such when you are developing your ideas. It may be that your topic has emerged from work experience or a very strong personal interest. Defining your project as a case study can help you to bring focus to your proposal.

Topics for case studies

Let's look at some examples.

Flood defences in Worcester
This project could be carried out by a student who had work experience with the Environment Agency. Perhaps she spent most of her time on placement working on proposals for improving flood defences in the Worcester area. She

goes to a potential supervisor to discuss the possibility of a research project 'to do with flood management'. The supervisor says, 'But a project can't just be reading books and writing a very long essay. How will you make this original?' They discuss her work experience and realise that her detailed knowledge of the schemes she worked on, and possibly access to some flood data, create the opportunity for an interesting case study. She checks with her former employer and they are happy to share data.

Conservation work on churches in Devon

This student has summer work experience with a building surveying firm in Devon who were doing some conservation work on churches. He was directly involved with three small schemes. He is interested in how historical, political and financial issues combine to force compromises in restoration work.

Methodology

Case-study research often involves a mixing up of methods – whatever suits the topic. The flood project might involve some flood modelling, some appraisal of existing defences, and study of scenarios involving different flood frequencies. The church restoration project might involve examination of the deterioration and remediation of relevant building materials, cost analysis, work with historical archives and interviews with parish council members. Both would be heavily supported by the literature; the case study creates a context for evaluation of the literature. With case studies like these your approaches must be defined; you can't make them up as you go along. You must be able to define the approach you plan to take and show that you have followed it. Where the work involves carrying out an appraisal of something, establish a formal grading system to provide objectivity, and apply it consistently.

General applicability

Your write-up must include some reference to how generally applicable your findings are. Some conclusions may be relevant to all flood defences or churches, not only those in Worcester or Devon. The strength of a case study is its focus, but the value is in its general relevance.

14.6   Writing a research proposal

This is an early stage in a research project. Students are commonly required to produce research proposals to guide them towards thinking out their research before they start. Professional researchers often have to write proposals as part of the competition for funding.

A student research proposal commonly has the following sections:

Aims
Literature
Methodology
Feasibility
Programme

Aims

A research proposal should set out the aims and objectives, and/or research questions, as discussed in 14.2. Remember that an aim is an overall direction and objectives are milestones on the way. Examples have been given in 13.3.

Literature

There must be some reference to the literature (significant work that has been carried out in the same field). In a proposal this may not be a full literature review, but must be enough to assure the reader that your research is not just a repeat of someone else's, and to show where your own research will fit.

Methodology

This is the overall approach to carrying out the research. It needs to be justified: you need to explain why you propose to approach the research this way and not another way. You will describe the data that will be generated, and explain how it will be presented and analysed.

Feasibility

Is it all do-able? What could go wrong, and what will you do if it does?

Programme

This is a timeline for the project from the start to submission of the report. A Gantt chart on an A4 sheet is usually ideal for this. You may find yourself thinking 'I'll never stick to this'. If that is because your durations are unrealistic, then change them; your programme must be realistic. If it's because you think things will go wrong – that's fine, things never go exactly as planned; you need to think through the details anyway; and you should keep the programme up to date as the project progresses.

14.7   Research reports

For many students the report on their research project is the crowning glory of the course. The report is what's left after the research is finished; it is the real evidence of your achievement.

Even though what you are writing may be called a dissertation or even a thesis, all the advice in Chapter 13 ('Reports') applies, it's just that the content now is fundamentally academic. This is a big test of your writing skills, so Chapter 4 should be useful too, including the advice on getting started with a writing task, and the tone and language to use.

The relevant earlier sections of this book are summarised below:

	Guidance	Example
Basics of written English	4.1	
Getting started	4.2	
Improving writing	4.3	
Appropriate style	4.4	
General report structure	13.2	Box 13.1
Contents	13.3	Box 13.1
Summary	13.3	Box 13.2
Aims and objectives	13.3	Box 13.3
Tables and figures	13.3	
Conclusions	13.3	Box 13.4

The key elements of a research report are:

Aims
Literature review
Methodology
Data presentation and analysis
Discussion and evaluation
Conclusions and recommendations

Aims

I'm using 'aims' to represent aim and objectives, or research questions, or hypothesis. You may need to introduce your topic before presenting your aims (as in Box 13.3). You will already have the latest version of your aims, the one that has guided you through your research. However, when you are finalising the research report you will have completed the research. The aims you express now are the ones you've achieved.

Literature review

This is now a detailed review of all relevant published work in the topic area. It's not a review like a film review, although if you think there is a weakness in a publication you can point it out, provided you make sure your comments are as objective as possible and can be well supported. In any case it is not a review that treats each publication separately. Rather it considers a general topic and indicates the contributions of the various studies. There is a 'language' to this. An example is given in Box 14.1. You will see that the citations are given in the standard Harvard form, introduced in 5.4.

Shelley *et al.* (2009) present an extensive study of part-time higher education in Scotland in a variety of subjects. They report a staff perception that part-time students in general are more committed and motivated than full-time students. They also found that 'part-timers were more likely to have a sense of where their occupational career might be leading and the role which education might play in this' (p159).

Bowen *et al.* (2009) surveyed 6000 part-time students, including built environment and engineering students. The results suggested that there were substantial personal economic benefits to be gained from engaging in part-time study, with the majority of respondents able to obtain better jobs and salaries at the end of their courses than at the beginning.

Many other studies of part-time students (in any subject) focus on the problems and challenges of part-time study without much consideration of the advantages. For example Wood and Cho (2014) and Wong *et al.* (2015), who studied part-time students in a range of subject areas in Hong Kong, consider the employment of 'coping mechanisms' by part-time students and identify the sacrifices that must be made. Dillon and King (2015), studying nursing students in the UK, concentrate on the high levels of stress experienced by part-time students. The paper makes no comment on any potential advantage of being a part-time student and does not, for example, probe areas in which part-time students might feel less stress than full-time students.

Edwards (2016) studied part-time civil-engineering students at the University of Devon. The study indicated a significantly higher level of performance overall by part-time students. All tended to have high levels of commitment to their studies which partly arose from the fact that most chose civil engineering as a *job* before they chose it as an academic subject. However, the consensus from the interviews was that the greatest advantage came from the skills, attitudes and motivation that part-time students had developed in the workplace.

A recent large-scale study of part-time students in employment in the UK (May *et al.*, 2010) covering a wide range of discipline areas (including engineering and built environment) has similarly identified the significance of the link between commitment to studying and career ambitions. For the vast majority of part-time students surveyed, the decision to study and choice of subject were firmly linked to career aims. Stone and Hoskins (2016) consider the perspective of employers, and found strong support for the combination of work experience and study experience by part-time students in the workforce.

Box 14.1 *Example of a literature review*

In your literature review you can also refer to what you might think of as 'established theory'. Some people present 'theory' in a separate section from 'literature' but there is no need. Literature just means what has already been published, and that includes theory.

Methodology

This will cover your overall approach and choice of methodology, and you must justify it. (If you wrote a research proposal, you can start with the relevant content in that.) This section will probably also cover the details of your procedure. Work out a logical order for setting this out: you don't have to follow the order in which you carried out the work, with all its rethinks and repeats. Here you are presenting the finished article, seen from a distance, not a personal history of developing ideas.

The form of language should be as discussed in 4.4. In some subjects it is normal to use the first person ('I carried out my study in three stages'), but in engineering and built environment it is not. Some academics feel strongly about this. It is simply a convention in the subject areas. As discussed in 4.4, you can use the 'passive voice' ('The study was carried out in three stages') but that can get a bit contrived. So avoid the first person, but don't feel you have to use the passive voice either ('The study had three stages').

You must provide enough detail for someone else to be able to replicate your study.

Data presentation and analysis

The relationship between words, tables and diagrams is key here, as discussed in 13.3. Have another look at the advice about using appendices. Don't think of the appendices as a kind of 'dumping ground'. The presentation, clarity and conciseness must be at the same standard as the rest of the report.

Discussion and evaluation

There is no standard approach to separating analysis and discussion. Sometimes analysis and discussion happen at the same time. But there will be a need for an overview of your analysis: what it all means when taken together. There must also be some evaluation, self-criticism if you like, of what you have done and how you have done it. If someone did replicate your study would they get the same results? This is a good way to start thinking about your evaluation. For a study of physical phenomena (as covered in 14.3) the answer should usually be yes. Are there any reasons why this might not happen? For a study of human factors (as covered in 14.4) it is much less likely that you could answer yes. Why?

Conclusions and recommendations

Students sometimes find this section hard, but it's very important that your reader is not left thinking 'that's interesting, but so what?'

The conclusions present the outcomes of your work, and what your study has added to existing knowledge. There is a sample (though not specifically related to research, and shorter than the conclusions that would be written for an extensive research project) in 13.3.

A research report often contains 'recommendations for further work', in other words if someone does more work in this area, what, in the light of your findings, would you recommend they study? There's an academic cycle going on here: previous studies (as described in the literature review) leading to your study; your study leading to further work.

15

Presentations

Let's keep this simple. What's happening when you give a presentation?

- You're speaking
- You're probably using visual aids of some sort, typically a computer-based presentation projected onto a screen, or you might have a poster or a model
- You're engaging with an audience
- You're communicating

These four go together. Speaking without visual support, something for the audience to look at while you're speaking, is 'giving a speech'. That's unusual for students as part of their course. So to start with there is a combination of the two media – speech and visual image. Also the visual image is often verbal (in the form of words) so your speaking can be supported by written words as well as pictures. Unless you are making the presentation to a camera, there is physical communication too through facial expressions and gestures, communicating enthusiasm and warmth, besides the content. There's a lot of communication going on – or there should be.

Here are my 10 guidelines for giving a presentation. Actually, at the risk of sounding bossy, I'm going to call them 'rules'.

1 Have something interesting to say

Often this isn't a problem – you're giving a presentation about your own project and you're confident that it's interesting. But what if you've been asked to give a presentation on something that you don't find interesting? Well, seriously, you must find some angle that you *do* find interesting, or do some research to find something interesting, or ask for a new topic. You really mustn't present if you don't have something interesting to say.

2   Think about how to say it

You've something interesting to say, so now you must think about an interesting way of presenting it. What would bring the subject to life, arouse the audience's curiosity? I'm not talking about making the presentation an interesting 'performance', I mean making the design and content of the presentation interesting.

3   Plan

How will the presentation be structured? How will it be sequenced? You may start experimenting with the presentation software at this point, or you may just want to make notes. It's a lot like planning something you are writing (as discussed in Chapter 4). Put yourself in the audience's position. They want to know first who you are and what you're going to talk about. OK then – that's the start of your presentation. You may then want to explain how your presentation is structured. How will you end? The worst sort of ending is when the last slide has passed, the screen is blank, and the speaker says 'Ah, er yes, that's it'. Plan your conclusion; you may want to repeat your main points. You need to give the audience the feeling at the end that the presentation has been worth listening to.

4   Don't read from notes

Eh? Where's this come from? It looks like this rule is a bit out of sequence – after all, we haven't finished preparing yet. But this is a rule we need to start thinking about now because it affects everything else. The annoying thing is that some people can read from notes very well – politicians, or television presenters, for example. Once I was asked to write a speech for the vice-chancellor of the university where I worked. It was amazing to hear my words, exactly as I had written them, spoken as if they were being made up on the spot. But that is a real skill, and it takes a lot of practice. Whenever I hear students giving presentations by reading from notes I find it boring at best, and usually embarrassing. Why? Because it is hard to read when you're nervous and so the delivery is very unsteady; because you can't look up, so there is no engagement with the audience; and because it is unnatural and so unless you're an actor or an experienced speech-maker you just sound like a dreary robot.

5   Prepare

Remember – you are speaking with support from visual material, and this can be in the form of words. The keywords or phrases that you put on the screen provide reinforcement to what you are saying to the audience, but they can also act as your prompts. The trick is to put just enough on the screen so that it reminds you of what you want to say. You need to think this out carefully.

6 Use technology well

Your presentation software, combined with images, video, animation, sound, can enhance your presentation, and if it does you are using technology well. The only time technology is being used badly is when it distracts from what you are saying. Presentations are not fundamentally about clever use of software: they are about interaction with your audience.

7 Practise

When the presentation is prepared, run through it plenty of times. You need to check that it takes the right amount of time. Practise in front of an audience if you can – maybe in the room where the actual presentation will take place. Also just run through it in your head. Don't try to memorise anything, just look at your materials and practise what you want to say. Once you've done enough practice you will know what you want to say when you look at each slide.

8 Don't fight nerves

'Nerves' are a form of excitement and mental readiness. If you know what you're doing, your presentation will be fine. If you have doubts or worries, it may not go well, but that's not 'nerves', it's poor preparation. Don't try to pretend you're not nervous, and if someone gives you terrible advice like 'well at least you should try to look confident', just tell them to go away (I'll leave the precise wording up to you).

9 Be yourself

So, no pretending. No pretending to be something you're not. Just be yourself and don't disguise anything. Show how you feel: nervous, excited, enthusiastic, proud.

10 Never forget your audience

Maybe this should be the first rule, not the last. A presentation is not about you, it is about your audience. Don't think 'will I say the right thing?', instead think 'will the audience understand and enjoy what they hear and see?' They must be able to hear you, of course, so speak clearly, and try not to speak too quickly. They must feel you are talking to them. They must be included. Look at your audience; look at every one of them. If there are lecturers and students, maybe visitors, present, don't look at one more than the other. Gauge your audience's reactions to what you are saying – you may be able to tell straightaway that you have moved to the next slide too quickly, for example, or that they are particularly interested in one of your images. But don't read too much into faces: don't be put off if they're not all nodding in enthusiastic agreement all the time. At the end say 'thank you' or something similar that acknowledges that an important human interaction is coming to an end.

So, what do you think of all that? Let me anticipate your responses.

I've got a lot of complicated stuff to cover – I'll never remember it all. I've got to read from my notes.

Firstly, there's no point in presenting material that is so complicated that an audience would not be able to understand it the first time they heard it. If you really think that is the case, the only option is to provide a handout – but that might be inappropriate, depending on the circumstances. Something complicated should be displayed on the screen for as long as necessary for the audience to take it in. To give them time, and to help them understand, you could talk them through the image in detail. It's a big mistake to show a complicated diagram, make a brief comment, and then move straight on to the next slide. Plan carefully.

I'd love to make it 'natural', sort of improvised, but I know I'll be so nervous I'll forget everything. I've got to read from my notes.

Your presentation doesn't need to sound improvised. It should sound like a well-prepared presentation given by a student who is naturally nervous about the event. It certainly won't sound natural if you read from notes. If you go 'er', 'um' a few times, so what? That's natural anyway. You won't 'forget everything' because all the prompts you need will be on your slides. That's what I meant by 'put just enough on the screen so that it reminds you of what you want to say'. If that means putting a bit more on the slides because you know you'll be nervous – that's fine. You won't mess it up because you'll have practised. That's why theatre groups or bands rehearse.

Thanks for the advice but English is not my first language. I've got to read from my notes.

I understand this makes things much harder. (Or perhaps I don't come near to understanding because I could not give a presentation in any other language, I admit it.) You can put the prompts you need on the slides; if you are worried about a particular section you can read some text from a slide (but not for the whole presentation); you can practice. Reading from your notes is not better, or easier, especially in a language that is not your first.

I hate presentations.

Most people say this. But if you ask them after a presentation which obviously went well how they felt about it, they always say something like 'yes, that was great!'. What people actually hate is giving bad presentations. Aren't you glad you've got this book!

It's worse when a presentation is a major assessment.

I get the impression that presentations are being used more and more as a form of assessment – not just to give practice as a preparation for professional life – but in place of other forms of assessment, even written exams. If a presentation is generating serious marks it obviously needs serious preparation. And of course this refers to the quality of the content in academic terms as well as to the competence of the delivery. But a presentation in place of an exam? That's not so bad really, is it?

It's worse still when it's a group presentation.

There are challenges in group work and these have been discussed in Chapter 11. In a group that is not communicating well, a presentation can be a problem. If you can't rely on your team members, then preparing and delivering a good presentation can be difficult. You can only solve the problem in the ways discussed in Chapter 11, by meeting and agreeing responsibilities. But in a group that is functioning and communicating well, it can be reassuring to be presenting as a team. You can pool ideas, share responsibilities, and support each other in delivery. Of course there are aspects of structure that need to be carefully thought through – the first speaker is introducing the presentation by the whole team, the last speaker is concluding it for the team. And remember, if all the team are 'on stage', don't talk and chat or fidget, or do anything that might distract your audience. Look at the presenter (your team mate) with interest. Look like a team.

Posters

Let's say you have to produce a poster (usually A2 or A3 size) on your project. Don't think 'how can I get all that information on one sheet?'; think instead 'how can I communicate the essence of the project with visual impact?' It's a challenge, but you simply mustn't sacrifice impact for detail. Create punchy summaries of your aims and conclusions, and use images wherever you can to communicate your approach and results.

Posters can be created using presentation, publishing or even word-processing software. I find all the best ideas come from my students, not from me.

We always invite someone from industry to judge the posters and award a prize. The prize invariably goes to a poster that seems surprisingly low in content but high in impact. (But if that surprises me I suppose it shows how little I know about designing posters!)

One last thing – don't forget to include your name. Having to scribble your name on your poster after you've pinned it up might spoil the effect.

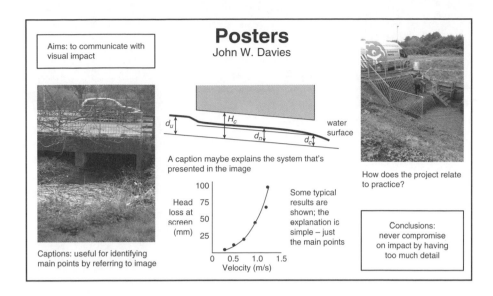

Posters
John W. Davies

Aims: to communicate with visual impact

A caption maybe explains the system that's presented in the image

Some typical results are shown; the explanation is simple – just the main points

Captions: useful for identifying main points by referring to image

How does the project relate to practice?

Conclusions: never compromise on impact by having too much detail

Professional activity

16
Work experience

Gaining relevant experience during your period at university is vital in today's competitive job market. It will complement and underpin your academic learning. There are many activities that can aid your employability:

- Year-long placements
- Vacation work experience
- Internships
- Volunteering
- Opportunities abroad
- Clubs, societies, committees
- Enterprise activities

The 'gold standard' as recognised by employers and universities is the certified year-long placement as part of the degree. Your own circumstances may dictate whether a year on placement is feasible or whether the next best thing, vacation work experience, is more suitable. However, everyone should be equipped with the information they need to make the decision for themselves. Here are the fundamental benefits to gaining work experience of any length.

> Your CV is going to look blank without a placement. ⚇ **FINAL-YEAR STUDENT**

> There are financial advantages, definitely. I'm going travelling for six months in South America on my savings from my placement. ⚇ **FINAL-YEAR STUDENT**

> [Another benefit of a placement] – a dissertation brief from your employer – much better than competing for a supervisor to do one of their topics. ⚇ **RECENT GRADUATE**

Benefits of work experience:

- Improves working knowledge and understanding of the discipline
- Likely to lead to a graduate job offer
- Can help reduce debt through payment of a salary for a year
- Can help provide career ideas and direction
- Gives experience of making job applications and presenting a CV
- Improves interview performance
- Can provide a final-year dissertation topic
- Can provide good industry contacts
- Can lead to final-year sponsorship
- Can lead to overseas travel for work
- Can contribute towards a professional qualification
- Development of 'work ethic' leading to enhanced academic performance back at university
- Renewed motivation for studying the subject from seeing it in context.

How does vacation experience compare with a year-long placement?

I understand that some people feel they don't have a choice: 'I'm 24 years old already; I need to finish my degree'. Also, when students have more than one vacation period with the same employer, it begins to build into the type of experience they would get from a year out. The relationship with an employer is a key thing. I know there are some excellent schemes in which employers offer students work experience throughout their degree – and I think that's great. Also in some disciplines, certainly architecture, professional practice is a more formalised component in the sequence of education and qualification.

But for situations where there is a choice, here are my slightly biased comparison points between a one-off vacation period and a year-long placement.

Most employers favour year-long placements for the reasons given in the last two rows of the table. Some employers simply don't provide vacation experience at all.

> *On placement, you learn so much better; it's more solid – you retain it.*
> 👤 **FINAL-YEAR STUDENT**

Vacation	Year-long
Doesn't extend the length of your degree	Does
It's a few months	It's a significant period – you develop
People will meet you; you will meet people	People will really get to know who you are; and you will really get to know them
Good for seeing a bit of the industry	Good for experiencing the industry (maybe more than one aspect)
When you've learned the job it'll be time to leave	You will learn the job and then have time to really develop
Even by the end you'll barely be worth your salary to your employers	You will be worth your salary

What will it be like on a placement?

You will get to do a real job and hopefully feel real achievement.

It's great putting theory into reality. 👤 **FINAL-YEAR STUDENT**

You will get a good idea of what a career in the discipline would really be like.

...helps you decide if you want to do it. 👤 **FINAL-YEAR STUDENT**

You'll learn a lot.

I liked learning about the commercial side – the relationship with the client. 👤 **FINAL-YEAR STUDENT**

There's a social side too.

There was a great atmosphere at work – you don't even think about that sort of thing before you start, but it's good. 👤 **FINAL-YEAR STUDENT**

It may be your first significant break in education since a very early age – a chance to recharge batteries.

I know that a lot of students worry that they'll 'get out of the studying habit'. I've never known this happen. It's the habits they pick up while they're out that count for most: not wasting time, knowing how to focus on a task, appreciating why it is worth doing well on the course.

What will be expected of you?

Some work experience contributes directly to your degree award. It may be that the degree is awarded as a 'Sandwich Degree', or the certificate may state that the degree has included a period of work experience. In other cases, the placement contributes credits to your degree. This could be in place of credit for academic modules, though often it is simply additional credit since a certain number of academic modules need to be passed anyway. You will need at some point to register for a year-out placement, if only to indicate or confirm that you will be away from the university for a year. Sometimes vacation experience has no formal link with the university.

Finding a placement is likely to be a joint effort by you and the university. The university will have contacts with employers, including those who have taken students in the past; also they may be approached by firms looking for placement students. They will make these details available to you. Of course all this will be explained when you say you are interested. You may also need to look for jobs yourself, especially if you are trying to find work in a particular location or technical field.

When opportunities arise you may need to make the application to the employer direct, or via the university (this will be made clear). Making effective applications is covered in Chapter 17.

When you are out on training, you are likely to be visited by a member of staff from the university to make sure everything is OK: that you are getting good experience, and that your employer is happy with the way you are settling in. You will also of course be able to contact the university at any time.

There will be some paperwork to complete early on, about contact details, health and safety arrangements, etc.

Typically you will be expected to keep records and provide reports on your experience.

It is very likely that you will be expected to keep a **diary**, recording briefly what you do every day, and significant events at work. This is to help you write your experience reports (below), and also because it is good professional practice to record all events that you might need to check on later: for example, when you received or passed on information, or encountered special problems. Make your diary readable and understandable to others: they may need to refer to it when you're not around. Ask your work colleagues about keeping a diary.

You may need to keep some form of **calendar** record so that the total period of experience can be certified.

And you will certainly need to write some form of **experience reports**. These will be the main formal record of your experience and will contribute to any assessment of the placement. The content is likely to be specified, but will include:

Basic facts – the period covered, employer, location, projects you have been involved in.

Description of experience – details of what you have actually done, with sketches, diagrams, bits of drawings, photos where appropriate; specific and technical, but with explanations when needed by someone not familiar with the work.

Appraisal – your personal judgement about what you are learning, how you are developing, your concerns and goals.

Your work placement may contribute to the experience you need to gain a professional qualification after graduation. That's obviously a good thing, and something to check on; it could affect how you keep your records.

You may be required to give a presentation on your placement, either while you are out, to an audience of work colleagues and university visitors, or when you resume your studies, to an audience of students themselves considering whether to take a placement.

As in all walks of life, it is exceptionally important that your employer has a good opinion of you. This will benefit you in future in the form of good references ('letters of recommendation' to possible employers), and, as happens regularly to successful placement students, offers of employment after graduation. That is a nice feeling in the final year – a job lined up if you want it!

This good opinion depends on the usual things: your enthusiasm, initiative, ability to take responsibility, etc. And it's worth remembering that we are not just talking about the opinion of your senior boss, but of all your work colleagues; it is always worth doing your best.

17

Job searching

First things

This is not the start. Well OK, it is the start of this chapter. But it is not the start of the process of getting a good placement or graduate job. If you've been reading previous chapters you'll know that I've been referring to 'keeping an eye on where it's all leading'. I hope you've been thinking about the prospect of getting a job before you reach the stage when you need to start looking for one. If you haven't, well of course I'm not going to say 'tough, you've left it too late', but you will have some catching up to do, not least in terms of thinking about how all the studying you've been doing might relate to a satisfying career.

> *It is never too early to start looking for employment.* 👤 **EMPLOYER**

There's a lot of advice/support out there. You are surrounded by advice on job applications, CV writing, interviews and so on. It comes from your university careers advice team, and probably also your faculty/school/department. There are lots of books (including this one), plenty of websites; there are the experiences of your friends and colleagues; there is encouragement from your lecturers. Even employers produce guidance on applications and selection procedures. You may find that surprising. Surely the point of an interview is to catch you out, expose your weaknesses; why would an employer want to give you clues in advance? But of course that's quite wrong. The purpose of an organisation's recruitment effort is to get the right people.

> *Know who is out there and what they are looking for.* 👤 **EMPLOYER**

It's not about pretending to be someone. This is the most important of the 'first things'. We've referred a few times in this book to the fact that being at university is about transition. You start (typically) as a school or college leaver; you end as a graduate, and, if it's what you want, you are ready to start a professional career. If, as a university student, you look back at yourself and your friends at school I'm sure you will see that you have changed. You are still the same person of course, but with different values, attitudes, outlooks, priorities. The same applies to the transition you are about to make. We discuss this more in the next chapter. It is likely that in a few years you will have a professional job, and you will be different. But that will be the 'new normal'. To present yourself to an employer now, you may have to imagine what you will be like when you have made that transition, but you don't have to pretend to be someone you're not. You may want to emphasise your strengths and come to terms with your weaknesses, but there is no pretending involved. After a few years at work, you may be sifting through CVs and conducting interviews yourself. Then *that* will be the new normal!

'*Be honest'* is advice about applications and interviews often given by employers themselves. 'Of course they would say that', the conspiracy theorist in you will say, 'they want to catch you out, find out the bad things about you'. But there are some compelling reasons for being honest. One is that it would be very embarrassing if a lie were exposed. But an even more important reason is this: how do you know what an employer really wants to hear anyway? Precisely what lies should you be telling? 'I can be lazy sometimes' – that's an example of something you might not want to say, not in those words anyway. But is it so bad? OK, 'I can be lazy sometimes' does sound like something out of a comedy book on how not to write a CV, but as a statement its problem is not its honesty but its naivety. Anyway, it doesn't really mean anything: there is no context, no way of judging what you are really saying. On a quiet Sunday morning you like staying in bed, not getting up to go to the gym or to read a book on business strategy. Great. We've all got to know how to switch off and relax. You get really fed up doing repetitive routine tasks. Great. Maybe that will spur you to think of a more efficient approach that would benefit everyone.

17.1　What do you want?

This really is the most important question. What motivates you? What is important to you? No one's saying the answer to these questions has to be 'work'. But your career is always going to be a significant element in your life, and you'll be much happier if you get things right.

> *Be clear in your own mind what motivates you. It might be to get recognition for your achievement, or to get a job to help other people, or to earn money to pay for your hobby. What matters is not what it is, but that you know what it is.* ⚇ **EMPLOYER**

I worked as a civil engineer for eight years, then I became a lecturer. I had enjoyed work in engineering, and I have always been slightly haunted by the saying 'those that can, do; those that can't, teach'. But I enjoyed lecturing straightaway and I have loved it ever since. I had some boring vacation jobs when I was a student and I know what it's like for time to drag at work. But I can honestly say that as a lecturer, not once in 30 years have I looked at my watch in the afternoon and thought 'oh no, it's still ages to home time'. I relish having a job I enjoy.

Think about what *you* want.

17.2   **Employability**

I hope you've noticed that employability has been a thread throughout this book. It is really where the book leads.

In Chapter 1	we started considering where your studies might lead.
In Chapter 2	we discussed how even while you get used to being a student you should keep the professional world in the back of your mind, and take opportunities to find out more.
In Chapter 7	we saw how the skills you develop on your course relate to the skills you need, and develop, as a professional, and how you will gain a professional qualification.
In Chapter 8	we considered ways of recording your development and achievements that will carry through to your career.
In Chapter 11	we considered how working on realistic projects prepares you for the profession.
In Chapter 16	we discussed the importance and significance of work experience.

And in Chapter 18 we will have a brief look at how a professional career develops.

What does 'employability' really mean? My definition is that it is the set of attributes that give you the best chance of getting a job you really want. This means taking responsibility for developing yourself and taking all opportunities for learning and gaining experience. Another more formal definition might be that it is the development of skills, abilities and personal attributes that enhance your capability to secure rewarding and satisfying outcomes in your economic, social and community life.

What do employers want? There's no simple answer of course. But here are some interesting quotes.

I'm looking for competent, honest, willing, reliable, professional, committed people with good interpersonal skills who can fit in with a team.

No employer wants a lazy, inefficient, difficult, disorganised, irresponsible, non-team-player – that's obvious – but the point is that if an employer gets the slightest hint of this at any stage [in the selection process] you will not get the job.

A good graduate is self-motivated, constantly looking to learn, keen to solve problems themselves and present their solution. 🙎 **EMPLOYERS**

17.3   CVs

The CV you have as a student is likely to be the most important two sides of A4 in your life. If submission of your CV is successful it will lead to you being shortlisted (selected for interview, or the next stage of the selection process) for a job. If you are eventually selected (for a placement, let's say), this first job will give you experience which will help to qualify you for whatever you want to do next – a graduate job, a better job, career advancement, promotion, etc. It all starts with your student CV!

There are many ways of communicating how important it is that you get your CV as good as it can be. There are plenty of horror stories of how many CVs (or applications generally) employers receive, and how many of them end up in the bin without being read in full. Sometimes I think students feel that these are just 'heartless employers' who don't realise they are throwing the aspirations of real young people in the bin. Surely not all employers can be that uncaring?

Employers want the best. The starting point is your CV. 🙎 **EMPLOYER**

You've got to look at it from their point of view. I have had to do plenty of shortlisting myself, mostly not for graduate positions or placements, but the process is the same. I remember once having 70 applications for two vacancies. Other work is pressing, time is limited, you can't go through 70 applications in great detail, you have to find a way of rejecting the less suitable ones quickly. After all, only two people are going to get a job at the end of it.

We look through CVs and written applications very carefully. Incorrect spelling, typing mistakes, poor structure, obvious mistakes with dates, these are all picked up and usually mean rejection. 🙎 **EMPLOYER**

Developing your CV

There are two aspects: content and presentation. Content covers your qualifications, your experience, your achievements, what you've done. You must include everything that's relevant. You've been preparing your CV all your life (by definition – that's what your CV is: your life so far). Presentation is also crucial. You should assume that one single misspelling, or inconsistent layout detail, will lead to rejection.

CV preparation is not a task, it is a continuous process. Here is a way of seeing it in stages:

1 *Think, make notes, expand your ideas.* If an employer was to know the best about you, what would that be? Genuinely – in every sense. Genuinely in the sense of honestly (as discussed above), but also genuinely from your point of view – what do *you* genuinely think is the best about you?
2 *Write a draft*, using all the guidance, templates, etc. you can find.
3 *Read it* carefully, cynically, looking out for phrases that are bogus, untrue, embarrassing or worthy of inclusion in a comedy book.
4 *Improve it.* Check again for any imperfection (spelling, layout detail).
5 *Show it to a careers adviser.* This support is there for you; it would be crazy not to make use of it.
6 *Act on their advice.*
7 *Make sure your CV is suited to each application.* 'Your CV' is not a fixed document; it may need to change for every use.

> *Get to know about the employer you are targeting.* ⚇ **EMPLOYER**

8 *Adjust it in response to each application experience.* Rejection or success – learn from every experience and improve your CV whenever you can.

> *A CV is a live document.* ⚇ **EMPLOYER**

CV for a placement

Two pages is the standard length for a student CV in the UK. When you are looking for a placement you may feel you don't have enough to fill two pages. You must give details of your education of course, and any work experience even if it's not relevant to your discipline. An employer will want to see what you've got to offer on the basis of what you've done so far. You will probably represent this as 'employability skills' or something similar. These must be genuinely about you – not a list of things you guess an employer would like to see.

Bethan Jackson is in the second year of the BEng course in Civil Engineering at the University of Devon. She is keen to get a year-out placement and is working on her CV. She is quite sporty, and takes on roles in relation to student interests, but her real passion is music. She performs with some other musicians and writes songs. Should this go on her CV? She thought not at first, but she mentioned it to a careers adviser and was urged to include it.

Don't forget your hobbies ... they complete you as a person. 🪨 **EMPLOYER**

She had several interviews. All the interviewers said something about the song-writing so they obviously noticed it. One asked a genuinely difficult question: 'What would you do if a song of yours became successful; would you keep working as a civil engineer?' Poor Bethan didn't know what to say, but the interviewer didn't seem to mind. 'I've worked with a lot of people who have strong interests and would pursue them more seriously if they got the chance; nothing wrong with that', he said.

The basic form of Bethan's CV at the time she was looking for a placement is on pp. 169–170. Remember: she would have tweaked this for each application to suit the organisation she was applying to.

CV for a graduate job
Hopefully when you're looking for a graduate position you will have more to put on your CV than when you're looking for a placement. As we have discussed in several places in this book, some work experience will be a significant advantage. If you don't have it, employers will wonder why. Employers are also interested in external interests, including charity work and volunteering, especially if this has involved team working.

Bethan's CV has been enhanced as a result of her work placement: see pp. 171–172.

17.4   Opportunities

You can find out about placement and graduate job opportunities in a number of ways:

- opportunities via the university – careers website, or via your department
- personal contact, from networking, previous work experience, or via your course
- company websites
- advertisements in professional magazines

The more local or personal the route, the better. When you apply to a vacancy advertised on the website of a large employer, you can be sure that your application will be one of hundreds. At the other extreme, employers sometimes contact lecturers in your department and say they are looking for a placement student or graduate. It does happen. You must pounce on these opportunities; you will be in competition with far fewer people and you have a ready-made direct contact. I think students don't always see this. Maybe a professional-looking advert is more attractive than a short note on the VLE or a note on a noticeboard. But play the odds! When you apply to the website of a large employer you are letting yourself in for that whole mass-selection thing – half the applications going straight in the bin, etc.

In the middle ground are the more 'official' opportunities on the university careers website. Go for these too; remember to keep checking for opportunities.

Best of all are your own personal contacts, most commonly resulting from work experience. Employers talk about placements as being 'year-long interviews'. Many successful placements lead on to graduate jobs. This is a very good reason for doing a placement.

17.5 Application forms and online applications

Most large employers require applications to be made online. Typically a series of questions will require you to reflect on previous experience and give examples of skills. Think carefully about your answers. If you can, view all the questions before you complete the application. Give yourself time to compose and refine your answers. You may be able to ask your careers adviser to check your proposed answers. You can also practise these difficult skills on dedicated websites. When you complete an application online you may also be asked to upload your CV.

In addition to an online application there may be several other steps in the recruitment process including psychometric testing, situational-judgement testing and assessment centres before you even get to interview! Many of these additional recruitment activities can be practised, specifically psychometric testing, so head for your careers service for resources. There are further details about some of these activities below.

Other employers require a paper-based application form. Many of the comments already made about CVs apply. Opportunities for giving emphasis to particular aspects of your background still exist, but there is less scope. If you have been asked to apply on an application form, don't send a CV instead, or fill in cursory details on the form and attach a CV. Instead, respond fully to the questions asked.

Commonly these forms can be downloaded as a file. This means that it is easier to correct mistakes, but it all still needs thought and care. Since every applicant's form will look similar, it's even more important that you think about how to make yours excellent, one that stands out.

As with a CV, be fussy. Take care over every detail.

Additional statement

Application forms tend to ask mainly for facts and then leave a space for 'Further information in support of application' or something like that.

Write what you think your potential employer should know about you, without exaggeration or padding. The form may say, 'continue on a separate sheet if necessary'. Don't be afraid to do this, but don't add waffle just so you need to use the extra sheet. Whenever you can, give specific examples of achievements. Take care with your use of English. The additional statement is a major test of communication skills (and will be seen that way by employers too).

Letter of application

You may need to send a covering letter with your CV or a paper-based application form. You should highlight some of the most relevant points in your CV, without being repetitive. In the letter, even more than the CV, you make your communication specifically for that organisation. Accuracy and appearance are important; it's a good idea to use the same font for the letter as for the CV.

17.6 Interviews

What kind of person are you really? That sounds like a horrible interview question, but in fact it sums up the purpose of the interview from the employer's point of view. Of course your qualifications and achievements matter, but the employer will already know these from your application.

I suppose you could say the interview is where it gets personal – especially as many selection procedures also involve other, more objective, tests and challenges (covered further in 17.7: assessment centres).

As we've already discussed, the *wrong* state of mind is that you have got to convince your interviewer(s) that you are something you are not. The answer to the question 'what kind of person are you really?' is worth thinking hard about, not only because it would help you answer such a question in the interview, but also because it will put you in the right state of mind. The only problem is that your interviewer wants to know what kind of person you would be if you worked for the organisation, whereas what's in your head is the kind of person you have been as a student. Maybe you should try out a few descriptions. Are you *determined*? You might think, 'well yes, I suppose I am. I hate giving up on things. I really worked hard on that design in year 2, there were times when I was supporting the whole group. It was very satisfying when it was all finished. But then I gave up cornet lessons after two months when I was 11'. Well, in relation to the challenges you can expect at work it sounds like you are determined. If your interviewer asks for examples you've got one ready: year 2 design.

Try some other characteristics: resourceful? outgoing? adventurous? good at reading people? understanding? patient? good with detail? Don't be hard on yourself – you can't be good at all these things. But you won't be bad at them all either!

Preparation for an interview

Part of your preparation for the interview will involve looking back at the material you collected when you made your application, and looking again at the application itself. You will know something about the organisation already, but before the interview find out more.

> *Go to our website; find out all you can about us. There's a big difference between candidates who really know something about us and those who don't.*
> ⚇ **EMPLOYER**

When preparing for the interview, in the same way as when working on the application, you should try to make full use of the careers guidance facilities at your institution. Your careers advisers will give you tips on interview technique and suggest further reading. They will have videos on being interviewed which you should find helpful. They may even put you through a mock interview.

As well as this you can practise answering questions in your imagination. You can do this anywhere – walking down the street, sitting on the bus, having a shower. You should imagine yourself as the interviewer in order to think of obvious questions.

The only really difficult questions are the ones you don't expect. Asking yourself testing questions can help to prepare you for the unexpected.

Test yourself

Here are some possible interview questions. How would you respond?

What do you know about this organisation?
What kind of person are you really?
Tell us a bit about yourself.
Why have you applied for this job?
What particularly interests you about it?
What makes you suitable for the job?
What are you seeking in any job?
What plans do you have for your career?

What excites you about a career in this field?
What talents/abilities will you bring to this organisation? Give examples.
What do you like about your degree studies?
What subjects are you good at/bad at?
Do you have any specialisms?
What challenges will this organisation face in the next two years?
Who else are you applying to? Which job would you accept if you were offered them all?

Many of the questions you will be asked will be based on your application. You should reread it thinking of likely questions.

Asking yourself questions will give you practice in composing answers. You should not try to memorise an answer to any question, because it will sound unnatural when you repeat it. However, using the 'STAR' technique for formatting your answer will help. STAR stands for Situation, Task, Action and Result. It is a technique that will help you relay the important information and achievements you want the interviewer to hear about in a coherent manner. Also, phrases you have composed during one of your imaginary answers will come back to you during the interview and can be worked into what you say. You should be ready to be asked at the start of the interview to give a brief introduction about yourself.

> *Be ready with examples of anything you think you might claim – that you are a good team player, or well organised, or whatever.* 🤵 **EMPLOYER**

The most effective form of interview practice is attending real interviews. This is certain to improve your technique, even if you might wish that you could get the perfect job first time. Try to be positive about rejection; you will always learn something. It is quite normal to contact the firm afterwards to ask for feedback on your interview performance.

The interview

You should arrive early, and eliminate any possibility of arriving flustered or disorganised. You must be dressed smartly; the best assumption is that you cannot be too smart for an interview.

You are likely to be interviewed by a panel. They will have done their homework, just like you have done yours. They will probably have your application marked up in detail, and will know what they want to ask you. But they will want you to do your best and there is no need to feel threatened or bothered by any of their questions. They will want you to like their organisation, because if they decide to offer you a job they will want you to accept it. If they ask a technical question it's almost certain to be easier than it sounds at first.

You will be asked at the end if you would like to ask any questions yourself. You should take up this opportunity, and use it to make sure you have found out everything important that you want to know. It is often wise not to ask about salary. (Your conditions of employment will be specified if you are offered the job.) You should make sure you fully understand the arrangements for training.

> *When we ask at the end of the interview 'have you got any questions for us?' and the candidate simply says 'no', I think 'surely there's something you want to find out more about?'* ⚬ **EMPLOYER**

17.7   Assessment centres

An assessment centre is a process (not a place). It involves a number of exercises that are designed to replicate the competencies and tasks of the role that the candidate is being assessed for. Employers use assessment centres to see how an applicant reacts in realistic work situations. Typical activities include:

- group exercises (business case analysis, model bridge-building, etc.)
- presentations
- problem-solving
- psychometric tests (including mathematical, verbal and personality tests)
- role plays (sometimes using actors)
- interviews.

The best advice is to be yourself and always try to contribute. Successful applicants are not necessarily those that push to be the 'leader' in team exercises, but those who show honesty, empathy, supportive behaviour, good social and communication skills, and an aptitude for team work.

The programme may include 'informal' sessions with all the interviewees together, at which you will be given information on the organisation or shown round, even taken for a meal. These are difficult times: you are obviously being assessed to some extent and being compared with the other applicants, but you should try to pace yourself. Being shown round and meeting people also allows you to think about whether you would be happy in the job. Remember that the whole thing is a two-way process.

> *Interact with the other people on the assessment centre, and co-operate with them. Be yourself.* ⚬ **EMPLOYER**

(For placement:)

Bethan Jackson

42 Clarence Road, Souweston, SU6 3SK
0787 569 0534
B.Jackson@devon.ac.uk

Personal Profile

Currently in the second year of the BEng in Civil Engineering at the University of Devon, I am very committed to finding work experience (year-out sandwich placement) in 2016–17. This will strengthen my understanding of the civil engineering profession and allow me to place what I am learning in my studies in a practical context. I have engaged enthusiastically with my course and have been elected as student representative by my peers. My work experience at Ryerson University, Toronto, involved working with minimal supervision and taking responsibility for the quality of my work.

Education

2014 to 2018	University of Devon Course: **BEng (Hons) Civil Engineering** due to graduate July 2018
	Modules (6, each 20 credits) currently studying in year 2 (2015–16): Structural Mechanics 2; Soil Mechanics; Hydraulics; Materials Civil Engineering Design; Construction Project Management
	Modules (6, each 20 credits) completed in year 1 (2014–15): Mathematics (68%); Structural Mechanics 1 (61%); Visualisation and Design (74%); Surveying and Setting Out (72%) Engineering Geology (66%); Civil Engineering and Communication (77%)
2012 to 2014	Somewhere College, Address **A-levels**: Mathematics (A), Physics (B), Geography (B)
2007 to 2012	Name of School, Address **GCSE** (7): Maths (A*), English (B), Physics (A), Chemistry (B), Geography (B), Spanish (C), History (C)

Employment

Summer 2015	**Lab technician (Civil Engineering), Ryerson University, Toronto, Canada** This entailed general laboratory maintenance and supporting researchers conducting experiments. On one project, I was given full responsibility for processing the results of a series of experiments, using quite complex procedures, making appropriate checks.

(continued)

Activities

Summer 2015	Travelled and worked in Canada and the USA.
Sport	Generally active. Main sport is athletics, representing school, college and university at 800 m and 1500 m.
Sports role	In 2015–16 I am **Fixtures Secretary for the Athletic Union** (all sports, appointed through election). I am responsible for coordinating block fixtures (several sports competing on the same day), and transport for all fixtures.
Course role	For 2014–16 I have been a **Student Rep**, one of two elected student representatives for my year on the Course Consultative Committee.
Leisure	Music – I am a singer-songwriter-guitarist and perform with others in pubs and clubs.

Relevant Employability Skills

Team working
A design task in year 1 and a more significant design project in year 2 required strong team working skills with formal evidence in the form of meeting minutes and a log. I scored above 70% for both. My sports role and course role (above) require effective people skills. I worked closely with researchers, academics and fellow laboratory staff in my vacation experience at Ryerson University.

Organisation
My role as Fixtures Secretary for the Athletic Union involves organisation of complex events and working with a wide variety of interested parties.

Technical Skills

Word processing (package name) – used extensively through school and university
Spreadsheets (package name) – used extensively through school and university
Presentation software (package name) – used extensively through school and university
CAD (package names) – 10-week course in year 1 and used extensively at university
Sketching – 10-week course in year 1; have maintained a sketch book throughout studies
Surveying and setting out – full module in year 1, with 3-day residential field trip

Other Details

Student member of the Institution of Civil Engineers and the Institution of Structural Engineers. Full, clean, driving licence held since 2014.

References

Dr H.N. Edge, Programme Manager	D. K. Chuckston, Laboratory Manager
Department of Built Environment	Department of Civil Engineering
University of Devon	Ryerson University
Town, Postcode	Toronto, Postal Code, Canada
Phone number	Phone number
Email address	Email address

(For graduate position:)

Bethan Jackson

42 Clarence Road, Souweston, SU6 3SK
0787 569 0534
B.Jackson@devon.ac.uk

Personal Profile

I will complete my BEng in Civil Engineering at the University of Devon in July 2018. I am seeking a job as a graduate civil engineer in the areas of water and environment. I have engaged enthusiastically with my course and have been elected as a student representative in each year of my studies. My 12-month work placement at Sanders Consulting Engineers has given me planning, design and construction experience relating to drainage, with significant responsibility for project management. My experience at Ryerson University, Toronto, involved challenging work with minimal supervision.

Education

2014 to 2018	University of Devon Course: **BEng (Hons) Civil Engineering**, graduating July 2018 *Year 3* (2017–18): Structural Mechanics 3; Environmental Engineering; Integrated Design; Research Project *Year 2* (2015–16): Structural Mechanics 2 (69%); Soil Mechanics (77%); Hydraulics (79%); Materials (65%); Civil Engineering Design (71%); Construction Project Management (62%) *Year 1* (2014–15): Mathematics (68%); Structural Mechanics 1 (61%); Visualisation and Design (74%); Surveying and Setting Out (72%) Engineering Geology (66%); Civil Engineering, Communication (77%)
2012 to 2014	Somewhere College, Address **A-levels**: Mathematics (A), Physics (B), Geography (B)
2007 to 2012	Name of School, Address **GCSE**: Maths (A*), English (B), Physics (A), Chemistry (B), Geography (B), Spanish (C), History (C)

Employment

2016 to 2017	**Sanders Consulting Engineers, Reading** My 12-month placement (June to June) gave me experience in design and project coordination. My projects involved adaptation of combined sewer overflows for Thames Water. I carried out assessments based on InfoWorks model information, developed schemes for improvement, sought approval and inclusion in the capital works programme, carried out the design and prepared contract drawings. I was also able to work with Thames Water's construction partner to oversee site works.
Summer 2015	**Lab technician (Civil Engineering), Ryerson University, Toronto, Canada** This entailed general laboratory maintenance and supporting researchers conducting experiments. On one project, I was given full responsibility for processing the results of a series of experiments, using quite complex procedures, making appropriate checks.

(continued)

Activities

Summer 2015	Travelled and worked in Canada and the USA.
Sport	Generally active. Main sport is Athletics, representing school, college and university at 800 m and 1500 m.
Sports role	In 2015–16 I was **Fixtures Secretary for the Athletic Union** (all sports, appointed through election). I am responsible for coordinating block fixtures (several sports competing on the same day), and transport for all fixtures.
Course role	For 2014–16 and 2017–18 I have been a **Student Rep**, one of two elected student representatives for my year on the Course Consultative Committee.
Leisure	Music – I am a singer-songwriter-guitarist and perform with others in pubs and clubs.

Relevant Employability Skills

Design
For Sanders Consulting Engineers (12-month placement) I developed engineering design skills, interpreting model data to optimise design.

Engineering process
At Sanders I also developed skills in scheme promotion and supervision of construction.

Team working
I have worked in a wide variety of team settings at Sanders, at Ryerson University and in university project work. My sports role and course role (above) require effective people skills.

Organisation
My role as Fixtures Secretary for the Athletic Union involved organisation of complex events and working with a wide variety of interested parties.

Technical Skills

Word processing (package name) – used extensively through school and university
Spreadsheets (package name) – used extensively through school and university
Presentation software (package name) – used extensively through school and university
CAD (package names) – 10-week course in year 1 and used extensively at university
Sketching – 10-week course in year 1; have maintained a sketch book throughout studies
Surveying and setting out – full module in year 1, with 3-day residential field trip

Other Details

Student member of the Institution of Civil Engineers and the Institution of Structural Engineers. Full, clean, driving licence held since 2014.

References

B. J. Hughes
Sanders Consulting Engineers
Halo House
Trentford Enterprise Park
Reading, Postcode
Phone number
Email address

Dr H. N. Edge, Programme Manager
Department of Built Environment
University of Devon
Town, Postcode
Phone number
Email address

18

A professional career

18.1 Early career

As you start your professional career, life changes dramatically from being a student. Some students dread it. But most graduates smile when they look back on their university days – yes, happy memories, but we move on!

People at different times of their lives simply see things differently. Let's imagine some reactions to key aspects of life.

Aspect	Student	Young professional
Freedom	I've got it, and I'll never have this much again.	OK, in theory you've got freedom as a student. But you want to do well on your course and you've got to do the work. So you put things off, then have to stay up all night to finish something. Is that really freedom? Plus I never had any money to do what I really wanted.
Exams	OK, accepted, not the high point of the student experience.	I have to become professionally qualified. There is an exam when you have the interview, but it's just essay-writing. Am I happy I don't have heavy exams every summer? Let me think...
Leisure	I've got flexibility. I can play sport on Wednesday afternoon. Stay up late (I mean late) some nights.	Life has to be more structured when you work for a living. Accepted – the working day is for work.

Aspect	Student	Young professional
Getting up in the morning	Sometimes I have important classes in the morning and I do (usually) make the effort to attend them. But having to get up early every morning is something I can do without.	Sometimes this is hard, I accept. But then there's that Friday-night feeling ...
Personal satisfaction	I get this from passing modules, getting the degree.	Work can be hard at times, but there's a lot of satisfaction – real achievement.
Team work	Can be stressful at university. Most people have at least one bad experience.	At work most people are part of a team and get to appreciate the mutual help and support that it gives. There's also a sense of belonging.

Your early career is still about transition, just like university. There's a lot of learning and personal development going on. But when you start your first professional job, don't let anyone tell you that you 'know nothing'. You may not have years of experience in the profession, but you have knowledge, ideas, energy, and a fresh and adaptable viewpoint. You are a valuable asset – that's why they employed you.

Apart from adapting to your new life, and enjoying having some money, you still need to pay some attention to the fact that you should acquire one more qualification – this time professional, not academic. The first few years of your career will entail initial professional development, as considered in Chapter 7, with the associated need to make records, as in Chapter 8. Most of the learning is on-the-job, and a lot of the record-keeping is routine, provided you keep things up-to-date. Then you'll go for your professional review, and you'll be a fully qualified professional!

18.2 Experience

If being an engineering or built environment student is mostly about gaining knowledge, understanding and skills, and being an early career professional is about professional development, being professionally qualified is about *experience*. Qualified professionals gain promotion, and move to better jobs, through their experience.

Another aspect of their professional life that becomes increasingly significant is management. At a certain level of responsibility their job is to manage: to manage staff, to manage resources, to manage projects and to manage time. Inevitably this means less detailed application of discipline-specific knowledge. A good day at work will be motivating staff, negotiating a change to the contract conditions, getting people together to make a key decision, or organising the team to keep on programme. Depending on the job, that may still mean technical decision-making. Even staff motivation is discipline-specific. Your young team will want their ideas understood and appreciated even if they sometimes have to be changed.

But you are also learning – that never stops. Why would you want it to? As we discussed in 8.3, CPD (continuing professional development) is an important part of being a professional.

References

Austin S., Rutherford U. and Davies J.W. (2011) Large-scale integrated project for built environment undergraduate students: a case study. In: Davies, J.W., de Graaff, E. and Kolmos, A. (eds) *PBL Across the Disciplines: Research into Best Practice*. Aalborg University Press, 222–232.

Blockley D. (2012) *Engineering – a Very Short Introduction*. Oxford University Press.

CIAT (2010) *Chartered Architectural Technologist, MCIAT – Professional and Occupational Performance (POP) Record, Competencies*. Available from www.ciat.org.uk [accessed July 2015]. Chartered Institute of Architectural Technologists.

Cottrell S. (2010) *Skills for Success: Personal Development and Employability*, 2nd edition. Palgrave.

Davies J.W. (2013) Using students with current industry experience to evaluate course delivery. *International Journal of Engineering Education*, 29 (5), 1199–1204.

Engineering Council (2014a) *The Accreditation of Higher Education Programmes, UK Standard for Professional Engineering competence*, 3rd edition. Engineering Council.

Engineering Council (2014b) *UK-SPEC: UK Standard for Professional Engineering Competence*, 3rd edition. Engineering Council.

Igarashi H., Tsang N., Wilson-Medhurst S. and Davies J.W. (2015) Activity-led learning environments in undergraduate and apprenticeship programmes. *International Joint Conference on the Learner in Engineering Education, IJCLEE 2015*, Mondragon University, Spain, July. ijclee2015.sched.org [accessed July 2015].

IMechE (2014) *Chartered and Incorporated Engineers Application Guidance, Version 2*. Institution of Mechanical Engineers.

Liengme, B.V. (2015) *A Guide to Microsoft Excel 2013 for Scientists and Engineers*. Academic Press.

QAA (2008) *The Framework for Higher Education Qualifications in England, Wales and Northern Ireland*. Quality Assurance Agency.

RICS (2015a) *Assessment of Professional Competence Pathway Guide – Building Surveying*. Available from www.rics.org/uk [accessed July 2015]. Royal Institution of Chartered Surveyors.

RICS (2015b) *Assessment of Professional Competence Pathway Guide – Quantity Surveying and Construction*. Available from www.rics.org/uk [accessed July 2015]. Royal Institution of Chartered Surveyors.

Rowntree D. (2000) *Statistics without Tears – an Introduction for Non-mathematicians*. Penguin Science.

Answers to 'Test yourself' exercises

Spelling

accommodation   argument   disastrous   environment   occurred
superseded   weir

Punctuation

1 What is the explanation?
2 These are examples of poor punctuation; they show what can go wrong. (Or a colon instead of a semicolon.)
3 Bad punctuation causes two main problems: ambiguity and poor readability. (Must be a colon, not a semicolon.)

Referencing and citing

Paragraph, something like:
What are students' attitudes to referencing? In the opinion of Huntsman (2013, p13) 'It seems clear that students understand that correct referencing is important (not least in terms of assessment) but they cannot help feeling it is trivial'. However, in a survey of 154 students at a university in the Midlands, 61% felt that correct referencing improved quality of writing, and 85% felt that correct referencing improved assessment of writing (Wiltshire and Ghent, 2012). The website of the organisation 'Student thirst' presents quotes from a survey on referencing, most respondents to which consider good referencing to be necessary but boring. One quote is: 'I honestly think lecturers care more about referencing than ideas' (Student thirst, 2014).

References

Huntsman E. (2013) *Students Can Write*, Apple Press.
Student thirst (2014) *About Referencing*. www.studentthirst.org.uk [accessed 12 December 2014].
Wiltshire A. and Ghent R. (2012) *Academic Writing: a Fairground Ride*, Report BB21, Higher Education Academy.

Estimating

1 volume = 0.02 × 10 × 5 = 1 m³.   mass = volume × density = 1000 kg.
2 0.1 kW × 100 h = 10 kW h   cost = 10 × 15 p = £1.50
3 about 1 m³/day or 0.2 m³/person/day   so consumption is relatively high

Accuracy

1 12.8 (it would be wrong to round to 12.85, then to 12.9; after all 12.849 is less than 12.850 so should be rounded down).
2 297 mm; divide by 7 and you get 42.42857. Using a ruler and pencil can you measure to the nearest 0.1 mm? – no. So perhaps we should say '42 mm'. But in that case the last strip would be 3 mm wider than the others. So I think you should say '42.4 mm'. That would make all the strips a bit more than 42 mm and hopefully close to being the same width.
3 2592000 (or 2.592 × 10⁶).
4 No! 0.004 × 2592000 = 10368p, or more than £103 – which would matter.

Visualisation

Yes.
Usually. No.
Yes.
A cylinder.

Interviews – only *you* can answer.

Index